Ibn Ezra
(Avraham Ben Meir Ibn Ezra)

The Beginning of Wisdom
RESHITH HOCHMA

Translated and Annotated by
Meira B. Epstein

Edited with Additional Annotations by
Robert Hand

Translation
©Copyright 1998
Meira B. Epstein

Additional Annotations
© Copyright 1998
Robert Hand

Second Printing August 2009

ISBN 978-0-9662266-4-5

Table of Contents

Foreword by Robert Hand i

Introduction by Meira Epstein iv
- On The Translation iv
- Technical Note: v
- The Hebrew Names of the Signs and the Planets v
- Books Consulted vii

The Beginning of Wisdom 1
- Introduction 1
- This Book Is Divided into Ten Chapters. 1
- Chapter One 4
- Chapter Two 14
  - Aries 14
  - Taurus 25
  - Gemini 31
  - Cancer 36
  - Leo 41
  - Virgo 46
  - Libra 50
  - Scorpio 55
  - Sagittarius 59
  - Capricorn 64
  - Aquarius 69
  - Pisces 73
  - The Superior [Fixed] Stars 77
  - These Are the Stars North of the Ecliptic 79
  - These Are the Stars South of the Ecliptic 80
- Chapter Three 82
  - The Wheel 83
  - The First House 85
  - The Second House 86
  - The Third House 87
  - The Fourth House 87
  - The Fifth House 87
  - The Sixth House 88
  - The Seventh House 88
  - The Eighth House 88
  - The Ninth House 89
  - The Tenth House 89

The Eleventh House 89
The Twelfth House 90
Chapter Four 91
Saturn 91
Jupiter 94
Mars 96
The Sun 98
Venus 100
Mercury 101
The Moon 103
Chapter Five 105
Positions of strength 105
Positions of weakness 108
The Harm of the Moon [Comes] in Eleven Ways 108
Chapter Six 110
Know That the Three Superior Planets 112
Venus and Mercury 114
The Moon 114
Chapter Seven 116
Application 117
Conjunction 117
Co-mixture 118
The Aspect 119
Separation 119
Solitary Motion (Void of Course) 120
Feral 120
Transfer [of Light 120
Collection [of Light] 121
Return of the Light 121
Conferring of Influence 121
Conferring of Rulership 121
Conferring of Nature 121
Conferring of Two Natures 122
Directness 122
Distortion 122
Prohibition 122
Returning of Good [Influence] 122
Returning of Harm[ful Infuence] 123
Cancellation 123
The Case of Three Planets in One Sign 123
Loss (Frustration) 123
Deprivation of Light 123

Pleasantness (Recovery) . . . 124
Recompense . . . 124
Reception . . . 124
Generosity . . . 125
Similitude . . . 125
Middle Position . . . 125
Authority . . . 125
Chapter Eight . . . 127
Chapter Nine . . . 139
The Lot of the Moon . . . 139
The Lot of the Sun . . . 140
The Lot of Saturn . . . 140
The Lot of Jupiter . . . 141
The Lot of Mars . . . 141
The Lot of Venus . . . 141
The Lot of Mercury . . . 141
The First House . . . 141
The Second House . . . 142
The Third House . . . 142
The Fourth House . . . 142
The Fifth House . . . 143
The Sixth House . . . 144
The Seventh House . . . 144
The Eighth House . . . 145
The Ninth House . . . 146
The Tenth House . . . 147
The Eleventh House . . . 148
The Twelfth House . . . 149
The Lots That Are by Themselves . . . 149
Lots, in Every Revolution, to Know All the Things That Will Become More Expensive or Cheap . . . 153
Chapter Ten . . . 156

Foreword by Robert Hand

The Beginning of Wisdom is one of the more interesting introductions to astrology of those texts written in the Middle Ages. In addition to being a very comprehensive basic introduction to the astrology of the Arabic Era, it contains much lore that is either not found or is not as clearly evident in other sources. I should like to use this foreword to give a brief summary of the highlights of this text as they occur in the various chapters.

- Chapter Two is an unusually thorough introduction to the various attributes of the signs of the zodiac. Particularly interesting, however, are the images associated with the decanates, or faces as they are known. It is very clear in Ibn Ezra's account that the ultimate origin of many of these images is the imagery associated with the constellations. This raises the question as to whether it is appropriate to use these in a tropical zodiacal system such as that used in the modern West (and in Ibn Ezra's time as well). Such a sidereal system is also the probable origin of such systems as "bright", "dark", "smokey", etc. degrees which makes their use in a tropical system very dubious as well.

- Chapter Three contains a fairly straightforward account of the houses. But it also contains material on the uses of the triplicity rulers of the house cusps. Much the same material is found in Bonatti, and Alchabitius, in both cases attributed to one Allendezgod. However, the presence of this material in Ibn Ezra indicates that this material is quite mainstream in medieval astrology even though it is largely unknown to modern astrology.

- Chapter Five is a particularly detailed and comprehensive account of the accidental dignities of the planets as they arise from their positions in their own orbits, in relation to the Sun (oriental and occidental), and their aspects with the other planets. His account of the planets increasing and decreasing in light, number, and calculation is paralleled in Abu Ma'shar's *Abbreviation of the*

Introduction to Astrology[1] but Ibn Ezra gives additional material as to its use in practical astrology.

- Chapter Six continues the kind of material introduced in Chapter Five and amplifies the practical aspects of the use of that material. His account of the phases of the planets with respect to the Sun is particularly comprehensive and is much more sophisticated than the distinction of oriental and occidental in the simple manner as it is found in later authors.

- Chapter Seven is a listing of the various circumstances of planets as they form aspects with each other in the signs. This is one of the more thorough accounts of what is called in Latinized Arabic *Alitifal* (spelling variations abound), the Persian-Arabic system of applications and separations of aspects. Also we have descriptions of various kinds of reception that is otherwise found only in Abu Ma'shar.

- Chapter Eight is an interesting collection of aphorisms including one that provides an interesting exception to the general rule of not allowing out-of-sign aspects that one generally finds in early medieval astrology. See aphorism #35.

- In Chapter Ten we have general rules on when to direct what and in what manner, but we also have something else. We have one of the few places outside of Abu Ma'shar where some of the basic doctrines of Persian mundane and historical astrology are discussed.[2] Unfortunately, it is only a taste. We wish Ibn Ezra had given us more.

A final comment: *The Beginning of Wisdom* is only an introduction to astrology. As Meira Epstein informs us in her introduction, this book is not only the beginning of wisdom, it is also the beginning of the quite complete textbook of astrology that one would find if one were to put

---

[1] A translation of which is available from ARHAT.

[2] See David Pingree, *The Thousands of Abu Ma'shar,* London: Warburg Institute, 1968, for a more complete but unfortunately still sketchy account of this system.

all of Ibn Ezra's writings together into one set. This is what we will do. There will be more in this series of texts.

Introduction by Meira Epstein

Rabbi Avraham Ibn Ezra (1089-1164 C.E, Spain), who was a multi-faceted Jewish scholar, wrote *Beginning of Wisdom* (*B.O.W.*) in the year 1148 as a basic text-book of astrology. This was his best known astrological text, which was copied numerous times over the centuries, and was also translated into other European languages. He also wrote a number of other astrological treatises, one of which is *The Book of Reasons* (*B.O.R.*) which I have already translated for Project Hindsight. *B.O.R.* is Ibn Ezra's additional commentary for *B.O.W.,* but it can be read by itself, on its own merit. Together they bring out his witty, analytical, and personal style. In my introduction to *B.O.R.* I included a broader description of his life and his other works.

I am often asked about Jewish astrology, especially in connection with Ibn Ezra. Ibn Ezra's writings are traditional Arabic astrology; it is neither Kabbalistic nor religious. Yet, at times his Jewishness shines through in little phrases, and in what I call his Talmudic style which is more apparent in *B.O.R.* Ibn Ezra also wrote commentaries on the Bible in which he included astrological references, and it is my hope to research and bring them out too at a later stage.

On The Translation

This translation is based on the Levy-Cantera Hebrew edition, John Hopkins University Press, 1939. That publication includes the Hebrew text, which they edited from several manuscripts, the translation of *Beginning of Wisdom* into French, 1273 by Hagin de Juif, and their own translation of the text into English. Until now the Levy-Cantera translation has been the only source for English-speaking readers who wished to study the astrological works of Ibn Ezra. Reading this translation made it obvious that the work deserves another attempt by an astrologer in order to bring this material to life, resolve certain phrases, streamline the astrological contents, and make it accessible to modern astrologers.

I wish to acknowledge one special source that I employed in doing this translation, *Star Names, Their Lore and Meaning* by Richard Hinckley Allen (1899). This is a book without which I could not have translated the second chapter on the signs and their various co-rising constellations, and I thank Diana Rosenberg for introducing me to it and re-kindling my interest in fixed stars. At first sight some of the descriptions of the signs and the constellations and the decanates were hopelessly unfamiliar, and only after thorough checking with Allen's book was I able to resolve most of them. I used mostly the Arabic and other early sources quoted in this book, assuming that Arabic science served as Ibn Ezra's source. Allen's book is extensively quoted in some of my footnotes and referred to as *S.N.*

In listing the names of the constellations and the stars I followed Allen's format of capitalization of the nouns only as it seems to me to make the reading more fluent. I applied the same pattern in places where Ibn Ezra's name can be identified or related to the known names of the stars/constellations. In order to distinguish between the tropical sign and the constellation I used Aries for the tropical sign and the Ram for constellation, Taurus vs. the Bull, Gemini vs. the Twins, etc.

Technical Note:

Text in square brackets [..] was added to make the sentence more fluent and at the same time preserve the original text or in cases where an astrological term was obviously missing. Double brackets [[..]] were used in cases where Levy-Cantera's edition has added text in single brackets [..]. Parentheses () were used as an on-the-spot notes instead of inserting a conventional footnote.

The Hebrew Names of the Signs and the Planets
The Signs

Aries, *Taleh* — טלה. Meaning: Lamb

Taurus, *Shor* — שור. Meaning: Bull. The root of the word is related to 'seeing' or 'looking' (*shur* שור), to 'song' (*shir* שיר) and to 'a line' or 'row' (*shura* שורה). Also, possibly to 'straight' (*yashar* ישר).

Gemini, *Te'omim* — תאומים. Meaning: Twins

Cancer, *Sartan* — סרטן. Meaning: The crab.

Leo, *Ari'eh* — אריה. Meaning: Lion.

Virgo, *Be'tula* — בתולה. Meaning: Virgin.

Libra, *Moznayim* — מאזנים. Meaning: Scales. The root is also of the word 'balance'.

Scorpio, *Akrav* — עקרב. Meaning: Scorpion.

Sagittarius, *Kashat* — קשת. Meaning: An Archer.

Capricorn, *Gdi* — גדי. Meaning: Goat, or rather the baby goat or kid.

Aquarius, *Dli* — דלי. Meaning: Bucket. The root is also of the verb 'to draw water'.

Pisces, *Dagim* — דגים. Meaning: Fish.

Planets

Saturn: *Shabtai* — שבתאי. The root is 'שבת' which is also the name for *Satur*day (*Shabat*) means 'to cease action' — a very proper description of Saturn's effect. In modern Hebrew the word for 'strike' or 'stop work' is *shvita* (שביתה) from that root. Always masculine.

Jupiter: *Tzedek* — צדק. The word means 'justice'. The root is also of words that mean 'to be right or righteous'. Always masculine.

Mars: *Ma'adim* — מאדים. The word means 'red', and the root 'אדמ' is most likely related to *dam* דם which means 'blood'. It is interesting to note the same root appearing in the word for 'person' (*adam* — אדם) and for 'earth' (*adama* — אדמה). Always masculine.

The Sun: *Shemesh* — שמש or, Hama חמה. Mostly feminine.

Venus: *Nogah* — נגה. The word means 'beautiful bright light'. Always feminine.

Mercury: *Kohav Hama* — כוכב חמה. The phrase means 'Star of the Sun'. Occasionally it is referred to as *Kohav* only. Always masculine.

The Moon: *Yare'akh* — ירח. The same word with different vowels (*Yerakh*) is used for 'month'. Masculine. Another name is *Levana* — לבנה, which means 'white'. Feminine.

Books Consulted

Ibn Ezra, *The Book of Reasons.* Translated by Meira B. Epstein, Published by Golden Hind Press, Berkeley Springs, WV For Project Hindsight, November 1994.

Claudius Ptolemy, *Almagest.* (2nd century C.E.) Translated by R. Catesby Taliaferro. *The Great Books of The Western World* series. Published by the Encyclopedia Britannica, Vol. 16.

Claudius Ptolemy, *Tetrabiblos.* (2nd century C.E.) Translated from the Greek paraphrase of Proclus by J. M. Ashmand, 1822. Printed in 1917 by W. Foulsham & Co., London. Republished in 1969 by Health Research, 70 Lafayette St., Mokelumne Hill, California 95245.

Firmicus Maternus, *Ancient Astrology, Theory and Practice.* (334 C.E.) Translated by Jean Rhys Bram of the Classics Department at Hunter College New York, April 1975. Published by Noyes Press, Park Ridge, New Jersey.

Al-Biruni, *The Book of Instruction In The Elements of The Art of Astrology.* (1029 C.E.) Translated by R. Ramsey Wright, University of Toronto Published 1934, London, Luzac & Co., 46 Great Russell St., Ballantrae Reprint, 10 George Street North, Brampton, Ontario, L6X 1R2, Canada.

William Lilly, *Christian Astrology.* Originally published by Partridge & Blunder, London 1647. Facsimile reproduction 1985 by Regulus Publications Co. Ltd. Third edition. ISBN 0-948472-00-6, 0-948472-01-4.

Richard Hinckley Allen, *Star Names, Their Lore and Meaning.* Republished from a 1963 edition by Dover Publications, Inc. 180 Varick St. New York. ISBN: 0-486-21079-0 Library of Congress Catalogue Card Number: 63-21808.

Vivian E. Robson, *The Fixed Stars & Constellations In Astrology.* First published 1923. The Aquarian Press, 1969. Wellingborough, Northamptonshire, UK ISBN 0 87728 232 3 (USA), 0 85030 463 6 (UK)

Robert Zoller, *The Lost key To prediction: The Arabic Parts In Astrology.* First published 1980 by Inner Tradition, 377 Park Avenue South, New York, NY 10016 ISBN 0-89281-013-0 (paperback).

Nicholas Campion, *The Book of World Horoscopes.* First edition 1988, Aquarius Press, ISBN 0-85030-527-6. Second edition 1995. Cinnabar Books, ISBN 1-898495-01-7.

The Beginning of Wisdom

[Introduction]

The beginning of wisdom is the fear of God, for it is the instruction.[1] For when a man does not follow his eyes and heart to fulfill his [worldly] desire, then wisdom will rest in him. Moreover, the fear of God will protect him from the laws and ordinances of the heavens all the days of his life, and when his soul separates from his body, it (the fear of God) will endow him with eternity and he shall live forever.[2] So, here I shall begin to interpret the laws of the heavens according to the rules as practiced by the ancients, generation after generation, and after I complete this book I shall compose a book of interpretation of the reasons,[3] and to God I shall pray for help, amen.

This Book Is Divided into Ten Chapters.

The first chapter [deals with] the form of the wheel, its parts, its signs and its images, the seven planets, their elevation, their strength, their motion, and their rulership.

---

[1] The word *hochma* (חכמה) means wisdom, but is also used for science and learning. As a matter of fact Ibn Ezra is slightly paraphrasing here a verse from the Bible, Proverbs 10:9: "The Fear of God is the beginning of knowledge and wisdom but fools despise instruction," which can also be read as". . . knowledge of wisdom. . ." 'Instruction' comes from the word *mu'sar* (מוסר) which Levy-Cantera have in the footnote from another manuscript while in the text they put *mosad* (מוסד) which means foundation. I chose *mu'sar* because it is typical of Ibn Ezra to introduce play on Biblical phrases, as I noticed already in *B.O.R.* This can also be revealing in terms of Ibn Ezra's personal philosophy, reconciling religious belief with astrological knowledge.

[2] Throughout the text we compare the Hebrew and its English translation to the the Latin translation of Peter of Abano. That English translation of that Latin edition begins as follows: "For then the fear of God will preserve him unharmed by the influences [of the stars] and fate up to the end of his life so that after the separation [of the soul] from the body [the fear of God] will ordain his soul an heir to the eternal life to come." [RH]

[3] His *Book of Reasons*.

The second chapter [deals with] the influence[1] of the signs, their ascension, their effect, the co-mixture of the [fixed] stars, and the images.

The third chapter [deals with] the aspects of the degrees, the influence of the quadrants of the wheel, and the twelve houses.

The fourth chapter [deals with] the nature of the seven planets,[2] their influence, and whatever they indicate for all that are created on Earth.

The fifth chapter [deals with] the planets, when their power increases, and when it diminishes [based on their house position and aspects].

The sixth chapter [deals with] the strength of the planets themselves [based on their orbital motion and position], and [also] according to their position before the Sun or after it.

The seventh chapter [deals with] the aspects of the planets and their conjunctions, their co-mixture, their separation, and the general rules regarding the mixture of their powers with one another, and all that these indicate.

The eighth chapter [deals with] the judgment of the planets in inquiries, nativities, and revolutions.[3]

The ninth chapter [deals with] the lots of the planets, the lots of the houses, and all other lots that astrologers have mentioned.

The tenth chapter [deals with] the orb of light of the seven planets, the way they are directed, and their translation over the degrees of the

---

[1] The word *ko'akh* (כח), which normally means 'power' or 'strength', is also used in the text in numerous places to denote astrological 'influence' or 'effect'. The translation was chosen based on the context.

[2] The word is *me'shartim* (משרתים) which means 'attendants'. This is the traditional reference to the function of the planets *vis-a-vis* the luminaries. It is consistently translated here as 'planets'.

[3] The actual chapter seems to refer to horary astrology only, but the same rules also apply to natal astrology as well as the annual revolution of the Sun at the Vernal Equinox.

wheel,[1] and what these indicate in general.

Every learned [person] who investigates this science can observe the motion of the seven planets, that they are faster in their motion than their motion in the superior wheel,[2] their motion in their [own] spheres against the superior stars which are in the wheel of the zodiac, and all the motions around the solid, which is the Earth, like the point inside the circle.[3]

Then one will realize that although these motions are even and direct, their effect will vary according to the regions.[4] This is known from the number of degrees of the wheel, its left (northern) and southern images (signs), and the knowledge of the seven planets, their nature — general and particular — and all their actions.

---

[1] Equating different lengths of time with a sign or a number of degrees for the purpose of prognostication.

[2] The primum mobile. [RH]

[3] The Latin says, "He is altogether wise who wishes to inquire what happens to the motion of the seven planets which are swift in motion around the degrees of the higher sphere. He will also perceive when the planets are near the superior stars which are in the circle of the signs. And [he will perceive] all of those motions which are about the center which is the Earth, whose manner of being is just like the [central] point in the circle." [RH]

[4] The word *gvulim* (גבולים) is used in other places to denote 'terms' or 'bounds' as in 'the bounds of the planets in the signs'.

Chapter One

The wheel is divided into 360 even parts called degrees, which is agreed upon by all the ancients and the later [astrologers] as this number has all the fractions (in whole numbers) from $^1/_2$ to $^1/_{10}$, except for the $^1/_7$. It also divides into twelve parts called signs, and each sign [divides into] thirty even degrees, and each degree [divides into] sixty primary minutes, and each of those [divides into] sixty seconds, and so on, until tens.

The names of the signs are Aries, Taurus, Gemini, Cancer, Leo, Virgo, Libra, Scorpio, Sagittarius, Capricorn, Aquarius, and Pisces. These are in the superior sphere which is the eighth,[1] [and] together with the left (northern) and the southern images they total 48 images, and the number of their stars is 1022 according to all the ancients, including Ptolemy. Out of these 346 stars are in[cluded] in the images of the signs.

The first image is Aries, and its stars are 13 which include the horns and the Belly.[2] In the sign of Taurus [there are] 33 stars and [the] Pleiades on its back. In Gemini there are 18 stars; in Cancer 9 stars; in Leo 27 stars; in Virgo 26; in Libra 8; in Scorpio 22; in Sagittarius 31; in Capricorn 28; in Aquarius 42; and in Pisces 34 [stars].

The Southern images are 15 and [the number of] their stars is 316. First[3] is the Sea Lion, which some call Bear,[4] and its stars are 22.

The second image is the Mighty One [with the] Dog,[5] and its stars

---

[1] Above the sphere of the planets.

[2] From *S.N.* p.79: "The eastern portion is inconspicuous, and astronomers have mapped others of its stars somewhat irregularly, carrying a horn into Pisces and a leg into Cetus."

[3] First with respect to the Vernal Equinox.

[4] Ptolemy calls it the Sea Monster. It lies south of the constellation of Pisces and Aries, and is now known as Cetus. It has the shape of a whale, and in Greek mythology was sent to devour Andromeda, but turned to stone at the sight of Medusa's head in the hand of Perseus. All three constellations (Andromeda, Perseus, and Medusa) lie nearby on the northern side of the celestial equator. I could not find any source for the alternative name 'the Bear'.

[5] This is clearly a reference to Orion. The text says *ha'kelev ha'gibbor* (הכלב הגבור), which translates literally as the Mighty (or strong) Dog, but *ha'gibbor* (הגבור) is also a noun — the Hero. It also fits Orion in the number of stars attributed to it in Ptolemy's sequence of the constellations, which Ibn

are 38. The third is the River, and its stars are 34. The fourth is the Hare, and its stars are 13. The fifth is the Great Dog, and its stars are 18. The sixth is the Little Dog where there are 2 stars. The seventh is the Ship, and her stars are 45. The eighth is the Animal,[1] and its stars are 25. The ninth is the Cup, and its stars are 7. The tenth is the Raven, and its stars are 7. The eleventh is the One Carrying the Lion[2] half of which is in the image of a man and half in the image of a horse,[3] and

---

Ezra follows.

From *S.N.* pp. 306-307: "The Syrians knew it as **Gabbārā;** the Arabians, as **Al Jabbār,** both signifying "the Giant" . . . the Arabian word gradually being turned into **Algebra, Algebaro** . . ."

"Hyde (1636-1703) quotes from an Arabian astronomer, **Al Babādur,** the Strong One, as popular term for the constellation."

The Arabic *Al Jabbar* and the Hebrew *Gibbor* sound the same, and are probably of the same root. *Gibbor* means mighty, courageous, hero, and Ibn Ezra whose sources were Arabic is clearly using here the Hebrew word for the Arabic name of the constellation.

In the Bible, Orion's name is *Ksil* (כסיל), which means 'a fool'. See *S.N.* p. 309.

The mention of the dog here, in my opinion, is only to further qualify Orion, who is traditionally associated with the dog constellation, Canis Major.
From *S.N.* p. 119: "The Arabian astronomers called it **Al Kalb Al Akbar**, the Greater Dog, so following the Latins, Chilmead [circa 1639, English writer on globes] writing it **Alcheleb Alachbar;** and Al Bīrūnī quoted their **Al Kalb Al Jabbār**, the Dog of the Giant, directly from the Greek conception of the figure." *Kalb* in Arabic means 'dog'.

[1] This constellation agrees in number of stars with Ptolemy's Water Snake, traditionally known as Hydra. [Additional by RH] The Latin edition of Peter of Abano also has a serpent here.

[2] This is reference to Lupus, called by Ptolemy the Wild Beast, near the hand of the Centaur.

According to *S.N.* p. 278 the designation Lupus is from the ". . . astrologers' erroneous translation of **Al Fahd,** the Arabian title for this constellation, their Leopard or Panther; . . . The Greek and Romans did not specially designate these stars and thought of them merely as a Wild Animal. . ." "The Arabians also called it **Al Asadah**, the Lioness, — found by Scaliger repeated on Turkish planisphere . . ."

[3] This eleventh constellation is the Centaur, Centaurus (Chiron).

its stars are 36. The twelfth is the Leopard, and its stars are 5.[1] The thirteenth is the Censer, and its stars are 7. The fourteenth is the Crown, and its stars are 13. The fifteenth is the Southern Fish, and its stars are 11.

The left (northern) images are 21, and their stars are 360. One is Ash and her sons,[2] and her stars are 7. The second is the Great Bear, and its stars are 17. The third is the Crocodile,[3] and its stars are 31. The fourth is the One with the Flame,[4] and its stars are 11. The fifth is the Barking Dog,[5] and its stars are 22. The sixth is the Northern Crown, and its stars are 8. The seventh is the one Walking On Its Knees,[6] and its stars are 28. The eighth is the Falling Eagle,[7] and its stars are 10. The

---

[1] To follow Ptolemy's sequence, this is again a reference to Lupus, or the Wild Beast, with one difference — Ptolemy assigns 19 stars to this constellation.

[2] *Ayish* (עיש) is the Hebrew name for the Little Bear (Ursa Minor). From *S.N.* p. 449: "The Arabians knew Ursa Minor as **Al Dubb al Asghar**, the lesser Bear, . . . although earlier it was more familiar to them as another **Bier;** and they called the three in the tail of our figure **Banāt al Na'ash al Sughrā**, the Daughters of the Lesser Bier."

Ibn Ezra is using the Arabic name *a'sh* (עש) for the Biblical *ayish* (עיש). In Job 38:32 : "התוציא מזרות בעתו ועיש על בניה תנחם (*Ha'totzi mazarot be'ito ve'aish al ba'ne'ha tan'hem*) (. . . *Ayish* and her sons. . .) The verb in this sentence '*ta'nhem*' from the root 'to guide' might be an allusion to the use of this constellations in ancient times as a guide for travelers.

[3] The Hebrew word is *ta'nin* (תנין). This constellation is known as the Dragon. See *S.N.* p. 205. [Additional by RH] The Latin edition of Peter of Abano also has dragon here.

[4] Ptolemy has here Cepheus, also with 11 stars. The Hebrew *ba'alat ha'lahav* (בעלת הלהב). [Additional by RH] The Latin edition of Peter of Abano has *Quarta, inflammatus stelle eius 11*, a reference to Cepheus.

[5] This one matches Ptolemy's the Ploughman in number of stars.

[6] Hercules.

[7] This constellation is known as Lyra or the Lyre, but Ibn Ezra apparently is quoting a different tradition. From *S.N.* pp. 281-283: "The occasional early title **Aquilaris** was from the fact the instrument which was often shown hanging from the claws of the Eagle also imagined in its stars." "The association of Lyra's stars with a bird perhaps originated from a conception of the figure current for millenniums in ancient India, — that of an **Eagle** or **Vulture;** and, in Akkadia, of the great storm-bird **Urakhga** before this was there identified with Corvus. But the Arabs' title, **Al Nasr Al Wāki**, — Chilmead's **Alvaka** —

ninth is the Hen,[1] and its stars are 17. The tenth is the [female] One Sitting on a Chair,[2] and its stars are 13. The eleventh is the One Carrying the Head of the Devil,[3] and its stars are 26. The twelfth is the Shepherd who has the Bridle in his Hand,[4] and its stars are 14. The

---

referring to the swooping Stone Eagle of the Desert, generally has been attributed to the configuration α, ε, ζ which show the bird with half-closed wings . . . Al Sufi, alone of extant Arabian authors, called it **Al Iwazz**, the Goose. Chrysococca write of it as . . . the Sitting Vulture, and it has been **Aquila Marina**, the Osprey, and **Falco Sylvestris**, the Wood Falcon. Its common title two centuries ago was **Aquila Cadens**, or **Vultur Cadens**, the Swooping Vulture, popularly translated the **Falling Grype,** and figured with upturned head bearing a lyre in its beak."

[1] This is the Swan, *Cygnus.* Ptolemy calls it the Bird, but Ibn Ezra, again is following another tradition. From *S.N.* pp. 192-193: "It was όρνις with other Greeks, by which was simply intended a **Bird** of some kind, more particularly a **Hen.**" "In Arabia . . . it usually was **Al Dajājah**, the Hen, and appears as such even with the Egyptian priest Manetho, about 300 B.C. . ."

[2] This one is Cassiopeia. From *S.N.* pp. 142-143: "Hyginus, writing the word **Cassiopeia,** described the figure as bound to her seat, and thus secured from falling out of it in going around the pole head downwards." "The Romans . . . also knew Cassiopeia as **Mulier Sedis,** the Woman of the Chair, or simply as **Sedes,** qualified by *regalis* or *regia;* . . ." "The Arabians called **Al Dhat al Kursiyy**, the Lady in the Chair. . ."

[3] This is Perseus, who is traditionally described as holding the head of Medusa, one of the Gorgons. From *S.N.* pp. 330, 332 states that the Arabians ". . . commonly called it **Hāmil Rā's al Ghūl**, the Bearer of the Demon's Head, which became **Almirazgual** in Moorish Spain." "**Algol,** the **Demon,** the **Demon Star,** and the **Blinking Demon,** from the Arabians' **Rā's al Ghūl**, the Demon's Head, is said to have been thus called from its rapid and wonderful variation; but I find no evidence of this, and that people probably took the title from Ptolemy. Al Ghul literally signifies a Mischief-maker, and the name still appears in the Ghoul of the *Arabian Nights*. . ." "The Hebrews knew Algol as **Rōsh ha Sātān**, Satan's Head."

[4] This is Ptolemy's the Charioteer, known as Auriga. Some descriptions include a goat and a kid in this constellation, whence the 'shepherd' came.

From *S.N.* p. 83-84: ". . . is shown as a young man with whip in the right hand, but without a chariot, the Goat being supported against the left shoulder, and the Kids on the wrist." "A sculpture from Nimroud is an almost exact representation of Auriga with the Goat carried on the left arm; while in Graeco-Babylonian times the constellation **Rukubi,** the Chariot, lay here nearly coincident with our Charioteer, perhaps running over into Taurus." ". . . the thirteenth is the One that Holds Back the Animal,[1] and its stars are 24.

Rein-holder, was transcribed **Heniochus** by Latin authors, and personified by Germanicus and others as **Erechtheus,** or more properly **Erichthonius,** son of Vulcan and Minerva, who, having inherited his father's lameness, found necessary some easy means of locomotion. This was secured by his invention of the four-horse chariot . . ."

S.N. pp. 86-88, about the star Capella in this constellation: "This has been known as **Capella,** the Little She Goat, since at least the times of Manilius, Ovid and Pliny, all of whom followed the . . . Greek of Aratos in terming it a *Signum pluviale* like its companions the Haedi, thus confirming its stormy character throughout classical days. . . . Pliny and Manilius treated it as a constellation by itself, also calling it **Capra, Caper, Hircus,** and by other hircine titles." "**Amalthea** came from the Cretan goat, the nurse of Jupiter . . ." "Others said that the star represented the Goat's horn broken off in play by the infant Jove and transferred to the heavens as **Cornu copiae**, the Horn of Plenty. . ." "The early Arabs called it **Al Rakib**, the Driver; for, lying far to the north, it was prominent in the evening sky before other stars become visible, and so apparently watching over them; . . . while, always an important star in the temple worship of the great Egyptian god Ptah, the Opener ['Ptah' sounds extremely close to the Hebrew verb 'to open'; MBE], it is supposed to have borne the name of that divinity, and probably was observed at its setting 1700 b.c. from his temple . . . at Karnak" "The Akkadian **Dil-gan I-ku**, the Messenger of Light, or **Dil-gan Babili**, the patron star of Babylon, is thought to have been Capella, known in Assyria as **I-ku,** the Leader, i.e., of the year; for, according to Sayce, in Akkadian times the commencement of the year was determined by the position of this star in relation to the moon at the vernal equinox. This was previous to 1730 B.C. when, during the preceding 2150 years, spring began when the sun entered the constellation of Taurus. In this connection, the star was known as the **Star of Marduk** [=Jupiter. MBE] . . . One cuneiform inscription, supposed to refer to our Capella, is rendered by Jensen **Askar,** the Tempest God. . . . In astrology Capella portends civic and military honors and wealth."

[1] This is Ptolemy's Serpentarius, also known as Ophiuchus, the Serpent Holder, or Struggling with the Serpent. From *S.N.* pp. 298-299: "Golius insisted that this sky figure represents a **Serpent-charmer,** one of the Psylli of Lybia, noted for their skill in curing the bites of poisonous serpents; and this would seem to be confirmed by the constellation's title **le Psylle** in Schjellerup's edition of Al Sufi's work.

"But the Serpent Holder was generally identified with Ἀσκληπιὸς, **Asclepios,** or **Aesculapius,** whom King James I described as 'a mediciner after made a god' with whose worship serpents were always associated as symbols

The fourteenth is the Animal,[1] and its stars are 18. The fifteenth is the Devil,[2] and its stars are 5. The sixteenth is the Flying Eagle,[3] and its stars are 9. The seventeenth is the Sea Fish, and its stars are 4.[4] The eighteenth is the Head of the Horse,[5] and its stars are 4. The nineteenth is the Winged Horse,[6] and its stars are 20. The twentieth is the Woman Who Has No Husband,[7] and its stars are 23. The twenty-first is the

---

of prudence, renovation, wisdom, and the power of discovering healing herbs. Educated by his father Apollo, or by the Centaur Chiron, Aescelapius was the earliest of his profession and the ship's surgeon of the Argo. When the famous voyage was over he became so skilled in practice that he even restored the dead to life. . ."

[1] Ptolemy calls it the Serpent of Serpentarius — the Snake held by Ophiuchus. The Hebrew *ha'haya* (החיה) means 'animal', but, as *S.N.* p. 374 tells us, the astronomers of Arabia called it *Al Hayyah*, the Snake. Obviously, Ibn Ezra was using the Arabic/Hebrew word, which posed no difficulty for his contemporaries.

[2] Ptolemy calls it the Arrow. Also known as the Archer, but *S.N.* p. 353 tells us of a beastly association that can justify the name 'Devil': "The formation of this constellation on the Euphrates undoubtedly preceded that of the larger figure, the Centaur Chiron; but the first recorded classic figuring was in Eratosthenes' description of it as **Satyr,** probably derived from the characteristics of the original Centaur, Hea-bani, and it so appeared on the more recent Farnese globe. But Manilius mentioned it, as in our modern style, *mixtus equo*, and with threatening look, very different from the mild look of the educated Chiron, the Centaur of the south."

[3] Known as Aquila, the Eagle. Ptolemy counts 6 stars in this constellation. From *S.N.* p. 57: "To the Arabians the classical figure became **Al 'Okāb**, probably their Black Eagle, . . . while their **Al Nasr al Tāir**, the Flying Eagle was confined to α, β, γ . . ."

[4] Ptolemy calls it the Dolphin and counts 10 stars. The '4' is probably a copyist error, due to similarity between *dalet* = 4 and *yod* = 10.

[5] Ptolemy calls it Forepart of Horse.

[6] Pegasus.

[7] This is Andromeda. Ibn Ezra apparently is using Arabic sources for this name. From *S.N.* p. 32: "In some editions of the *Alfonsine Tables* and *Almagest* she is **Alamac** . . . and **Andromeda,** described as *Mulier qui non vidit maritum*, evidently from Al Biruni, this reappearing in Bayer's *Carēns Omnino viro*. Ali Aben Reduan (Haly), the Latin translator of the Arabian commentary on the *Tetrabiblos,* had **Asnade,** which is in the Berlin Codex reads **Ansnade** *et est mulierquae non habet vivum maritum*; these changed by manifold transcription

Triangle, and its stars are 4.

This completes the number of stars mentioned [above], which is 1022. The ancients divided them into 6 levels [of magnitude], for any star whose light is great is called of the first honor,[1] and any one of lesser light is of the second honor, and so they decrease until the sixth honor of which there is no lesser one. Of the first honor there are 15 stars; of the second honor there are 48; of the third there are 208; of the fourth there are 474; of the fifth there are 217; and of the sixth there are 49; and 3 stars are dark, like clouds.

The seven planets are Saturn, Jupiter, Mars, Sun, Venus, Mercury [and] the Moon. The most elevated is Saturn, for it is in the seventh sphere from the Earth, followed by Jupiter, until the Moon in the first sphere which is closest to the Earth.

The signs are divided into four groups according to the four elements, and, therefore, three signs are of the same elements. Aries, Leo, and Sagittarius are hot and dry, like the fire element; Taurus, Virgo, and Capricorn are cold and dry, like the earth element; Gemini, Libra, and Aquarius are hot and moist, like the air element; Cancer Scorpio, and Pisces are cold and moist, like the water element.

Saturn is cold and dry, Jupiter is hot and moist, Mars is hot and dry, and burning; the Sun is hot and dry; and Venus is cold and moist. Mercury is variable for its nature is like that of the star it is with. The Moon is cold and moist. Of these seven [planets] some are masculine, and some are feminine. Some are day planets and some are night planets, some are benefic and some are malefic. There are two luminaries, two benefics, two malefics, and one of a nature that varies with every nature (of the other planets). One of the lights is the Sun, which is masculine and its power is by day. The second is feminine and that is the Moon whose power is by night. One of the benefics is masculine and of the day planets, and that is Jupiter. The second is feminine and of the night planets, and that is Venus. One of the malefics is masculine and of the day planets, and that is Saturn. The second is feminine and of the night planets, and that is Mars.[2] Mercury

---

from **Alarmalah,** the Widow, applied by the Arabians to Andromeda; . . ."

[1] The Latin of Peter of Abano uses *ordo* meaning 'rank'.

[2] The following table shows the arrangement just described:

is mixed for it is variable, at times masculine, and at times feminine, at times of the day planets, and at times of the night planets, at times benefic, and at times malefic, according to the planet it is with in conjunction or aspect as I shall explain in the proper place.

The nature of the five planets as well as the Moon varies according to the distance from the Sun as I shall explain later. They have another variation of ascent and descent for at times they are near the Earth, and at times they are distant from her, and at times they are on the left (north) side, and at times on the right (south) side. [This occurs[1]] when the planet is in the elevation of the solid sphere, which is not like the solid [sphere] of the Earth, and then it is in its place of elevation from the Earth. It is the opposite when the planet is in its lowest position.[2]

	Masculine/Day		Feminine/Night
Benefic	Jupiter		Venus
Malefic	Saturn		Mars
Lights	Sun		Moon
Neutral		Mercury	

[Additional by RH] It is implicit in the doctrine of sect in which Mars is a nocturnal planet, that Mars might also be feminine. It is also implicit in the fact that Mars prefers the moisture (a feminine quality) of Scorpio to the dryness of Aries. However, Ibn Ezra is one of the few who have come out directly and classified Mars as feminine as well as nocturnal.

[1] The following is an explanation for the "ascending and descending" variation mentioned in the previous sentence, and it refers to the motion of the planet in its eccentric deferent circle

[2] The Latin of the Peter of Abano text translates as follows: "Also there are likewise differences in their ascent and descent. For sometimes they are near the Earth but sometimes they are distant from her. Likewise they are sometimes on the right and sometimes on the left. But when a planet is in the highest degree of its eccentric (the center of which is not the same as the center of the Earth), it stands then in the place of its auge most remote from the Earth. And conversely when it will be in the place nearest to the Earth, it stands then in the opposition of its auge." The auge of a planet is just what this passage describes, the point in the eccentric most distant from the Earth. For a more detailed explanation of this see page 22, note 1.

Their Hanging[1] is the joining place of the wheel of the planet, similar to the wheel of the zodiac with its inclined wheel.[2] The Head of the Hanging (north node) is the beginning of the left (north) and the Tail is the beginning of the right (south). When a planet, or the Moon, is with [its] Head or with [its] Tail, then it is on the ecliptic, but in other positions it will have northern or southern latitude relative to its distance from the two points.

The planets have rulership in the wheel [of the zodiac]. There is rulership by domicile, by exaltation, by triplicity, by bound, and by face. Rulership by domicile [equals] five fortitudes. Rulership by exaltation equals four, by triplicity three, by bound two, and by face one.[3] When a planet is in its domicile, it has power in the whole sign, and so it is in the sign of its exaltation; yet, at the [specific] degree of exaltation it has more power than in the rest of the sign.[4]

In this book I will mention to you all that was agreed upon by the ancient Babylonians, the Persians, the Hindus, and the Greeks, whose chief was Ptolemy. I will mention the 'nines' and the 'twelves',[5] and the bright and dark degrees, and the empty ones, the masculine and the feminine, and the pitted degrees, the degrees that increase grace and

---

[1] The node. The following is an explanation for the latter part of the above sentence dealing with latitude of the planets.

[2] And the Latin adds, "And its 'dragon' [as in Dragon's Head and Tail] is established in the place of the increase of the deferent circle of the planet corresponding to the circle of the signs in its place of increase." [RH]

[3] This is one of the earliest accounts of the method of weighting dignities. [RH]

[4] The Sun is exalted at 19° Aries, the Moon at 3° Taurus, the North Node at 3° Gemini, Mercury at 15° Virgo, Venus at 27° Pisces, Mars at 28° Capricorn, Jupiter at 15° Cancer, and Saturn at 21° Libra. These traditional attributes have no known source. Ibn Ezra refers here to the question whether the power of exaltation is limited to the specific degree or prevails in the whole sign. In Hindu astrology the exaltation degree has an orb, so to speak, so the closer the planet is to its degree of exaltation the better.

[5] Dividing of a sign by 9 or by 12, associated with sequences of the signs and their rulers. [Additional by RH] These are the *navamsas* and *dwadasamsas* which appear to have come into Arabic Era astrology from India, although the *dwadasamsa* also appears in the Greek tradition as one of the two kinds of *dodekatēmoria.*

honor, and the place of the numerous [fixed] stars that are in the zodiac, their longitude and latitude, and the nature of the greater ones among them, until my book is complete; and you will not need another book beside it for the beginning of this wisdom.

Chapter Two[1]

Chapter two [deals with] the signs, their ascension, the effects, and the nature of the prominent stars in them.

Know that the signs are twelve, six of which are to the left (north) of the equating line,[2] and they are from the beginning of Aries to the end of Virgo, and the southern ones are from the beginning of Libra to the end of Pisces.

Aries is a fire sign, masculine, diurnal, oriental, of the warm season. It is changeable,[3] since the season changes, and at its beginning the day equals night; and then the day begins to grow longer and the night grows shorter. Its [temporal] hours increase from the even [equatorial] ones.[4] It is of short ascension for it rises at all locations in less [time] than the equator,[5] and it is crooked.[6] It indicates the temperate heat element that increases and in it is [the time] when people travel.[7] It is of the good and agreeable signs. It indicates thunder and lightning. Its

---

[1] Heading not in original. [RH]

[2] The celestial equator. [RH]

[3] Cardinal. Changeable = Moveable.

[4] Temporal hours are the result of dividing the time from sunrise to sunset by 12. In the spring and summer these hours are longer than the even ones, which are measured at equal intervals of 15° along the equator.

[5] This is true in the northern hemisphere at least. In the southern hemisphere Aries is of course a sign of long ascension. [RH]

[6] This is a reference to Aries' short ascension. This must not be confused with the previous statement about the seasonal length of daytime hours during Aries month. Here we are talking about Aries rising on the Ascendant in the horoscope, at any location, any day of the year. Short or long ascension results from the obliquity of the equator with regard to the ecliptic. In the equatorial regions Aries rises in 1:52 hours. At latitude 40° Aries rises in 1:20 hours, etc. (No wonder Aries people are impatient. . .) The ideal would be 2 hours per sign. [Additional by RH] In most of the Arabic Era texts and the Latin ones derived from them, the word 'crooked' is used to denote short ascension because the signs do not rise upright in the east but instead rise obliquely with respect to the east horizon, 'crooked' as opposed to 'upright'.

[7] An interesting connection to travel is found with regard to α Arietis, Hamal, in *S.N.* p. 81: "Hamal lies but little north of the ecliptic, and is much used in navigation in connection with lunar observations."

front[1] [part] is rainy and windy. Its middle is temperate, and its end [has] heat and pernicious atmosphere. If it is [the] northern [part], it produces heat with moisture, and when it is [the] southern, it cools.[2]

It is a four-legged figure with hoofs, and its limbs are severed because of its crooked ascension in the wheel. It has two colors[3] and two forms,[4] and it has half a voice. Its is the center of the east and the easterly wind. Its share of the four elements is the fire and its heat. To it belongs the blood. Of the flavors anything that is sweet;[5] of the colors, ruddiness and anything that is saffron; of the animals, sheep and all [animals] that have hoofs; and of the metals, gold, silver, iron, and copper.

In its share are the third region,[6] and the land of Babylonia, and

---

[1] The Hebrew word *te'hilato* (תחילתו) may also mean 'its beginning', which can be understood as pertaining to seasonal time — the first part of the Aries month — but when checking with Ptolemy it is clearly about the area of the sign/constellation being investigated.

[2] This is a reference to the influence of the sign on the weather similar to Ptolemy in his *Tetrabiblos*, Book II, chapter XII where it is under the heading "The particular natures of signs by which the different constitutions of the atmosphere are produced." In a preceding paragraph Ptolemy explains that these attributes are observed from the new or full Moon nearest to the ingress of the Sun into the cardinal signs at the change of the season.

[3] The word is *eyna'im* (עיניים) which means 'eyes', but *a'yin* (עין), the singular form, also means color or hue, which is more fitting here, since Aries is associated with the colors white and red. Lilly, p. 86: ". . . white mixed with red . . ." See also *S.N.*, p. 79.

[4] *S.N.* p. 78: "Aries generally has been figured as reclining with reverted head admiring his own golden fleece. . . but in the 'Albumasar' of 1489 he is standing erect, and some early artists showed him running towards the west. . ."

[5] In other sources all of the fire signs are equated to the bitter taste. However, Aries is the beginning of the spring which is the time of the Air element, i.e., the hot and moist. Perhaps Ibn Ezra assigns the sweet taste to Aries based on its being the beginning of spring rather than assigning the taste on the basis of its being a fire sign. [RH]

[6] The third clima. Each of the signs is assigned one of the seven climes (clima) based on the ruler of the sign. Thus signs ruled by Saturn get the first clima, those ruled by Jupiter the second, those by Mars the third, Leo gets the fourth, signs ruled by Venus get the fifth, signs ruled by Mercury the sixth, Cancer gets the seventh. However, there seem to be some errors in this text.

Persia, and Adharbaijan,[1] and Palestine.

Of places [his are] the fields and pastures of sheep, and fiery places, and [the refuge place] of thieves, and all houses covered with a roof.[2]

Virgo should get the sixth clima like Gemini, and Aquarius should get the first like Capricorn. Instead both get the second. This is almost certainly an error since the other signs are all consistent with the basic principle.

The climata were belts of latitude related to the length of the longest day. The following is the scheme of the seven climata as described by Al-Biruni. [RH]

The Climates	Longest Day		Latitude	
	Hrs.	Min.	°	'
Equator	12	00	00	00
Beginning of the first climate	12	45	12	39
Middle	13	00	16	39
Beginning of the second climate	13	15	20	27
Middle	13	30	24	13
Beginning of the third climate	13	45	27	28
Middle	14	00	30	39
Beginning of the fourth climate	14	15	33	37
Middle	14	30	36	21
Beginning of the fifth climate	14	45	38	54
Middle	15	00	41	14
Beginning of the sixth climate	15	15	43	23
Middle	15	30	45	22
Beginning of the seventh climate	15	45	47	11
Middle	16	00	48	52
End of the seventh	16	15	50	25

[1] The text has *Ardayan* or *Ardyan* (ארדיין) and the Hebrew footnote has *Adryan*. Al-Biruni has *Adharbaijan*, which sounds close enough.

[2] William Lilly, p. 93: ". . ..a place of refuge for thieves (as some unfrequented place), in houses, the covering, ceiling or plaistering of it, a stable of small beats, lands newly taken in, or newly plowed, or where brick have been burned or lime."

Fire is the highest element in the cosmic order, and therefore indicates the highest place in the house, the roof. Air, Water & Earth follow in this sequence.

According to Ptolemy its are the houses of prayer and the place of judges.[1] The ancients said that of the letters it has Alef (א) and Nun (נ);[2]

This kind of information is used in horary astrology to locate the thing or person inquired about.

[1] In the *Tetrabiblos,* Book II chapter 8, on the subject of possible effects of eclipses through the different signs, Ptolemy says: "The equinoctial signs [Aries & Libra. MBE] further indicate the circumstances liable to happen in ecclesiastical concerns, and in religious matters."

Firmicus Maternus, in his *Mathesis*, Book II, chapter X, says about Aries: "This sign is called 'Crios' by the Greeks for the reason that when the Sun is in this sign, it judges, so to speak, between night and day, which in Greek is *crinein* (to judge); and because the Sun, located in that sign between summer and winter, is in turn judged." The translator, Jean Rhys Bram, comments: "Crios is the Greek for Aries, the Ram. It is Firmicus' idea to derive it from *crinein,* which means "to judge" in Greek." Robert Schmidt, the Greek translator of the Golden Hind Press Series of texts, confirms this comment.

The concept of 'judgment' with regard to Aries is also mentioned in Ibn Ezra's *Book of Reasons* (p. 13) where he says "The houses of worship of God. Aries and Libra . . . signify justice, and from justice, houses of worship". This is a very interesting addition to the attributes of Aries not found in later astrological texts. Libra does signify balance and fairness, which have to do with the justice process. Aries and Libra mark the Equinoxes, where day equals night, hence 'equality' — another dimension of justice.

[2] *Alef,* as the first letter of the Hebrew alphabet, signifies 'first', but it is interesting to note that the ancient form of this letter in the Semitic writing represented the head of a bull, and in the Bible the word *Aluf* (אלוף) also refers to bull or ox, as well as to persons in very dignified positions. As the Semitic alphabet developed from the Egyptian hieroglyphs in the early part of the second millennium B.C.E., I believe we have here an indication for the evolution of the tropical zodiac out of the sidereal one. In that era the constellation of Taurus was the constellation opening the year at the Vernal Equinox, and symbolically was represented by the sacred Egyptian bull Apis, as the incarnation of the dead god Osiris. As the Ram, Aries, took over from Taurus, due to the precession, it also acquired the title of 'the first'. As for the other letter, Nun, I will offer my own theory: 'Nun' in old Semitic languages meant 'fish', and as the equinoctial point moved down to the Fishes/Pisces, it is possible the Aries received this title too (See *S.N.* p. 337-338).

Attributing letters of the Hebrew alphabet to astrological elements is known only in the teachings of the Jewish mysticism of Kabbalah, originating in *Se'fer Ye'tsira* (*The Book of Creation*) which, according to tradition, is attributed to the Patriarch Abraham, but modern scholars trace its beginning to early medieval

its years are 15, its months are 15, and the days are $37^1/_2$, and the hours are 4.

In the first face there ascends[1] the figure of a radiant woman,[2] and the tail of the Sea-Fish that resembles a serpent,[3] and the head of the Triangle, and the form of an ox.[4] Hindu people say that there is the head [in the] form of a dog with a candle in its left hand and a key in its right hand. Banbakha, their learned one, said that in the first face there ascends a Moor; his eyes are black, and his eyebrows are straight, and he is of the giants [race]. He is self-laudatory and is wrapped in a large white cloak with a rope girdle on it; he is irascible and stands on his feet.[5] According to Ptolemy there ascends the back of the Woman

---

period. Given the astrological elements and other ideas contained in it, Kabbalistic thinking can be traced to antiquity in the Near-East, in the era when astrology flourished. Ibn Ezra deviates from those in *Se'fer Ye'tsira*, where Aries gets *heh* (ה). I do not have any supporting documents for Ibn Ezra's different attribution and the above explanation is my own theory, assuming that Ibn Ezra had access to other sources, or may be even created his own. I cannot tell whether this was intentional or not, but numerologically, *heh* (5) equals *nun* (50).

[1] Possibly co-rising with that face.

[2] This is probably Andromeda, judging from the placement, even though I could not find a reference to Andromeda as 'radiant'.

[3] This could be a reference to the eastern fish in Pisces, but more likely, it Cetus, the Sea Monster. *S.N.* p.161: ". . .in all delineations it has been a strange and ferocious marine creature, in later times associated with the story of Andromeda, and at first, perhaps, was the Euphratean Tiāmat, of which other forms were Draco, Hydra and Serpens;. . ."

[4] The connection to an ox image is found in *S.N.* p. 415, where he says that Triangulum had been associated with the island of Sicily, shaped as a triangle, "at times identified with the mythological Thrinakia of the *Odyssey,* the pasture-grounds of the Oxen of the Sun, that Gower called Mella's Holy Ox-land."

[5] *S.N.* p. 36, describing Beta Andromeda: ". . . the Arabic Mi'zar, a Girdle or Waist-cloth." "the corresponding sieu (Chinese lunar mansion)...the Man Striding, or the Striding Legs. . ." This star is shared with the northern fish of Pisces that lies very close to Andromeda, and is 'Al Risha', the Cord that unites the two fishes (*S.N.* pp. 338, 342). Hence, may be, the Girdle.

[Additional by RH] Compare Ibn Ezra's description of this face to that of the *Picatrix.* "the form of a black man, restless and great in body having red eyes and holding a cutting axe in his hand, girded about with white garment; and there is great worth in this face. And this is a face of strength, high rank,

who is Seated on a Chair, and her knees, and her left hand, and half of the back of the Woman who Has No Husband with her Pudenda[1] and the Train of her Garment,[2] and the second Fish, and the Flaxen Cord.

In the second face there ascends fishes, and the middle of the Triangle, and half of the Animal, and a Woman with a comb in her hair, and a bronze armor, and the Head of the Devil. Hindu people say that there ascends the figure of a woman draped in clothes and a mantle, and she has one leg, and she has the form of a Horse.[3] According to

---

and wealth without diffidence. And this is its form."

This same basic imagery is also found in Agrippa's *De Occulta Philosophia*, Book II, chapter xxxvii. ". . . in the second face ascendeth a form of a woman, outwardly cloathed with a red garment, and under it a white, spreading abroad over her feet, and this image causeth nobleness, height of a Kingdom, and greatness of dominion . . ."

[1] The word used in the Hebrew is 'fears' which was quite problematic. Levy-Cantera translated it as ". . . the spinster with her uterus. . ." I have seen the word *fears* (פחדים) used also for male genitals before, but not for the female counterparts. I have translated it as 'pudenda' on the assumption that the use of the word 'fears' in Hebrew was a euphemism much like the word pudenda is in Latin and English, that is, "that about which there must be modesty." [Additional by RH] The Latin of Peter of Abano which translates as follows confirms this reading: "and there ascends a half of the back of a woman who is wholly lacking in a husband. With this are lifted up her private parts and genitalia [covered with?] with fringed clothing."

[2] This is ξ Andromeda. *S.N.* p. 38: ". . . **Al Dhail,** the Train of the Garment. . ."

[3] *S.N.* p. 35: Regarding α Andromeda — "**Alpheratz, Alpherat,** and **Sirrah** are from the Arabians' **Al Surrat al Faras**, the Horse's Navel, as this star formerly was associated with Pegasus, whence it was transferred to the Woman's hair; and someone has strangely called it **Umbilicus Andromedae.** But in all the late Arabian astronomy taken from Ptolemy it was described as **Al Rās al Mar'ah al Musalsalah**, the Head of the Woman in Chains" "In the Hindu lunar zodiac this star, with α, β, and γ Pegasi, — the Great Square — constituted the double *nakshatra* — the 24th and 25th — **Pūrva** and **Uttara Bhādrapadās**, the Former and the Latter Beautiful, or Auspicious Feet; also given as **Proshthapadās,** Foot Stool Feet." The 'leg' mentioned here could be a Cassiopeia element. *S.N.* p. 145: "As a stellar figure in Egypt Renouf identified it with the **Leg,** thus mentioned in the *Book of the Dead* . . . 'Hail, leg of the northern sky in the large visible basin.' "

Ptolemy there ascends the Woman Sitting on a Chair, and the head of the One holding the Devil['s head] and his right hand, and the Train of the Garment of the Woman without a Husband and her feet, and the Triangle, and the head of the Ram and its horns, and the rest of the Flaxen Cord.

In the third face there ascends a young man sitting on a chair with a cover over him, and in his hand icons, and a man bowing his head, and he is crying to the Lord. There also ascends the Belly of the Fish and its head, and the end of the Triangle, and the second half of the Animal. Hindu sages say that there ascends a yellow man whose hair is reddish, and he is irascible and contentious, and in his hand are bracelets of wood and a wand, and his clothes are red, and he is a blacksmith, and he desires to do good but he cannot.[1] According to Ptolemy there ascends the Bearer of the Devil's Head, and the body of the Ram.

One born in [this sign] will have a medium body, long face and large eyes, and he looks towards the ground mostly. His neck is thick, his ears are weak, he has a lot of curly hair, and weak legs. His speech is pleasant, he talks too much, and likes to eat a lot. He is irascible, and pursues justice, and his voice is not strong. If the Moon is in this sign or her Lot[2] is with one of the malefic stars, it indicates skin disease and

---

[Additional by RH] Again from the *Picatrix:* "And there ascends in the second face of Aries a woman dressed in green clothes and lacking in one leg. And this is a face of high rank, nobility, worth, and kingship. And this is its form."

This again similar to Agrippa *op.cit.* ". . . in the second face ascendeth a form of a woman, outwardly cloathed with a red garment, and under it a white, spreading abroad over her feet, and this image causeth nobleness, height of a Kingdom, and greatness of dominion . . ."

[1] *Picatrix:* "In the third face of Aries there ascends an inquiet man holding in his hands a gold bracelet, and dressed in red clothes, who desires to do good and is not able. And this face is one of subtlety, and of subtle professions, and of new things and instruments, and the like. And this is its form."

Agrippa: ". . . in the third face ariseth the figure of a white man, pale, with reddish hair, and cloathed with a red garment, who carrying on the one hand a golden Bracelet, and holding forth a wooden staff, is restless, and like one in wrath, because he cannot perform what he would. This image bestoweth wit, meekness, joy and beauty." [RH]

[2] The Part (or Lot) of Fortune.

leprosy, deafness, and baldness.

One born in the first face will be yellow, with a narrow chest, and little flesh. He has a mark on his left leg and his left arm. His friends are many, and he abhors evil.

One born in the second face will be dark, with a handsome face, and his body will be medium. He is short-tempered but does not hold a grudge, with moral integrity, and his enemies are many.

One born in the third face will be yellow and saffron, and will avoid people.

In the share of this sign is the head and the face, and the pupil of the eye, and the ears. Its ailments [are the kind] when one suddenly falls but does not feel [anything],[1] and the aches of the ears, and the nostrils, and the teeth, and the eyes, and the blemishes that look like sores.[2]

According to the Egyptian astrologers the ailment of Saturn, [when found] in it, is [in] the chest,[3] Jupiter's is in the heart, Mars' is in the head, the Sun's is in the genitalia, Venus's is in the feet, Mercury's is in the legs, and the Moon's is in the knees.

Of people, in the share of this sign are kings who are just and generous, warriors, and fire, and killing and blood, and travelers on the road.

---

[1] Epilepsy.

[2] Leprosy? [Additional by RH] Schoener lists both leprosy and impetigo as being indicated by Aries.

[3] When a planet is in its own domicile, it is equivalent to being in its first house corresponding to Aries. As we know, the signs rule parts of the body, beginning with the first (Aries) ruling the head, the second ruling the neck, the third ruling the arms and shoulders, the fourth ruling the chest, etc. Therefore, when Saturn is in Aries, it rules the chest since Aries is fourth from Capricorn. This scheme applies to the rest of the planets mentioned in this paragraph and will appear in the description of the other signs.

It is the domicile of Mars, the exaltation of the Sun at the 19th degree, the fall of Saturn at the 21st degree, the detriment of Venus, and the inferior position of Mercury at 25 degrees at the present time in the year 4908.[1]

---

[1] This is a reference to Mercury's geocentric orbit from the point of view of Ptolemaic astronomy. Each planet is conceived of as being carried on two basic circles, the main one called the deferent, and a smaller one called an epicycle. The following diagram is one that deals with the planets in general not Mercury in particular. But this diagram is drawn in a manner common to many medieval texts.

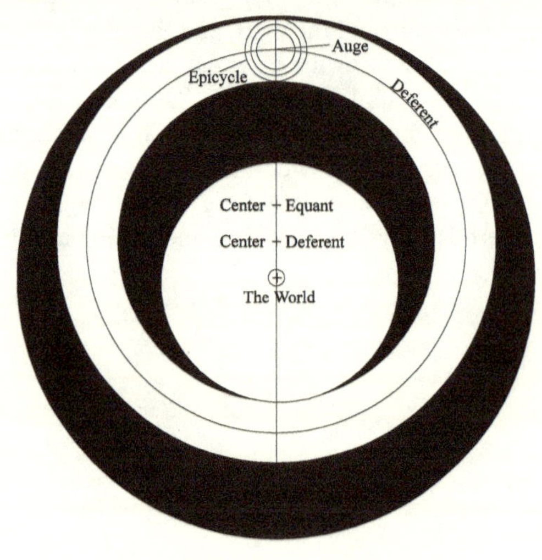

The circles bounded by dark areas were actually conceived by many authors as being solid and crystalline with them all rolling around within each other.

At the center of the outermost circle is the Earth "The World." Note that the deferent circle is not centered upon the Earth. It is therefore called an eccentric deferent or simply eccentric. The acual center of the deferent is the point marked "Center-Deferent." Then there isa third center marked "Center-Equant." This is a point such that movement on the deferent with respect to this center is at a constant rate. The center of the epicycle is carried around the deferent while the planet in turn is carried around upon the epicycle. The movement of the point at the center of the epicycle is **roughly** equivalent to the heliocentric position of the planet, at least in terms of its dynamic significance. As one can see from the diagram, a planet is carried nearer or farther away from the Earth by two motions. First of all different parts of the deferent are at different distances from the Earth, and second, the motion of epicycle (which causes retrogradation) also carries the planet toward and away from the Earth. The part of the *deferent* that is farthest from the Earth is called the 'auge' an Anglicized version of the Arabic *awj.* It is often translated as 'apogee' but this is not appropriate because it does not distinguish this farthest point from ones created by the epicycle. Also exactly opposite the auge

The ruler of [its] triplicity by day is the Sun, followed by Jupiter, and [by night it is Jupiter] followed by the Sun. Their co-ruler by day and by night is Saturn.

The first face according to the Egyptian astrologers and most of the other gentile astrologers belongs to Mars, the second to the Sun and the third to Jupiter, but according to the Hindu astrologers the first [face] belongs to Mars, the second to the Sun, and the third to Venus.[1]

These are the bounds according to the Egyptian and the Babylonian[2] astrologers: to Jupiter [are allotted] 6 degrees, to Venus 6, to Mercury 8, to Mars 5, and to Saturn 5.

According to Ptolemy to Jupiter [are alloted] 6 [degrees] first; [. .

---

is the point closest to the Earth, a 'perigee' of sorts, but this usage has the same problem as the use of the word apogee for the auge.

These auges were considered to move quite slowly and consequently their position could be given as being reasonably accurate for a considerable length of time. In this particular case what is given is the opposition of the auge of Mercury. The auge is at 25°♎. Therefore, the lowest point or the opposition of the auge is at 25°♈. A list of the auges from Al-Biruni follows:

Saturn	6°♐ 48'
Jupiter	16°♍ 43'
Mars	8°♌ 33'
Sun	24°♊ 32'
Venus	24°♊ 39'
Mercury	23°♎ 43'

The Moon has no given auge because it moves very rapidly. These values are a bit different from Ibn Ezra's because of the difference in date. Al-Biruni is 1029 C.E. where as Ibn Ezra is 1148 C.E.

For more information on this subject and related issues in geocentric astronomy see the notes to pages 14-18 in our edition of Abu Ma'shar's *Abbreviation of the Introduction to Astrology*.

[1] This is the Chaldean order of the planets which follows the their elevation in the spheres: Saturn, Jupiter, Mars, Sun, Venus, Mercury, and Moon. [Additional by RH] Actually, as far as which type of face rulership comes from whom, Ibn Ezra has it backwards here. The second type of face rulership (♂, ☉, ♀) according to the Chaldean order that Ibn Ezra mentions is found in ancient Western sources whereas the first type which gives the rulerships according to the rulers of the three signs in the triplicity (♂ - ♈, ☉ - ♌, ♃ - ♐) along with Aries (not to be confused with triplicity rulers) is the Hindu type. It has also become the most common modern Western system for decan rulership.

[2] These are not to be confused with the Chaldean bounds given by Ptolemy in Book I of the *Tetrabiblos*. [RH]

. lacuna in Hebrew text . . .][1]

. . .the second to Venus, the third to Mercury, the fourth to the Moon, the fifth to the Sun, the sixth to Mercury, the seventh to Venus, the eigth to Mars, and the ninth to Jupiter. One-ninth of the sign is three degrees and one-third. And all these ninths are of the nature of the sign and the ruler of the sign.

The influence [of] the 12 [divisions of the sign:] The 1st belongs to Mars, the 2nd to Venus, the 3rd to Mercury, the 4th to the Moon, the 5th to the Sun, the 6th to Mercury, the 7th to Venus, the 8th to Mars, the 9th to Jupiter, the 10th and 11th to Saturn, and the 12th to Jupiter. But Enoch and the ancients allotted the power of the 12 [divisions] in a different way, for they said that the 1st degree is of the nature of the sign itself, the 2nd [degree] is of the nature of the sign that follows it, [so that] the 13th degree and the 25th degree return to the sign itself.[2]

Hindu astrologers say that the first 3 degrees of a sign are medium — neither bright nor dark —, followed by 5 dark degrees, then 8 medium degrees, then 4 bright ones, then 4 dark ones, then 5 bright ones, and 1 dark degree.[3]

The first 7 degrees are male, then 2 are female, then 6 are male, then 7 are female, then 8 are male.[4]

---

[1] From the Latin we have the following: "Also according to Ptolemy, first, 6 [degrees] are of Jupiter; second, 8 [are] of Venus; third, 7 [are] of Mercury; fourth, 5 [are] of Mars, and fifth, 4 are of Saturn.

"And the first novenary is of Mars, the 2nd of Venus . . ." The text "second, 8 [are] of Venus" to "2nd of Venus" was obviously truncated by a copyists error. [RH]

[2] This appears to be another form of *monomoiria* in which each degree corresponds to a sign. See Paulus chapters 5 and 32. [RH]

[3] Al-Biruni's scheme for the bright and dark degrees is somewhat different. Regarding the interpretation of this attribute, he says on p. 270: "The distinction drawn between luminous and dark degrees is . . . not founded on any system and consequently recourse must be had to the subjoined (given) table. Astrologers, however, use it for making decisions as to colors, good and evil, strength and weakness, joy and sorrow, difficulty and ease. But no two books are to be found which agree on this matter, nor are they likely to found."

[4] Ibn Ezra's degrees agree with those given in Al-Biruni's table (p. 269). However, Al-Biruni expresses great doubt as to the value of this classification, since it is not based on any reasonable foundation. He mentions controversies regarding it, and other classifications in use. [Additional by RH] There is

These are the pits of the stars — the 6th, 11th, 17th, and the 23rd degrees. The degree that bestows grace and honor is 19.[1]

Of the great stars of the first magnitude there are in it the star called the End of the River, which at the present time is at 17 degrees of Aries, with latitude of 13½ south,[2] and its nature is like the co-mixture of Jupiter and Venus.

Taurus is of the earth signs, female, of the nocturnal signs, and of the warm season. It is a fixed [sign] and its [temporal] hours increase from the even ones. It is of short ascension and it is crooked. It is of the benevolent and refined signs, and it indicates procreation. On the whole, it is a mixture of hot and cold, and its moisture is greater than its dryness. Its front part indicates winds and darkness, its middle is cold and moist, and [the hind part] indicates thunder, and fire, and lightning, and meteors. Its left (north) side is temperate and the right side (south) is scorching. It is a four-legged figure with hoofs [but] its limbs are missing,[3] and it has half a voice.

In its share is the right side of the south and the southerly wind, and its nature is cold and dry [but] somewhat tempered.

Its is depression, somewhat lessened.[4] Of the temperaments it is

---

another system of masculine and feminine degrees which is found in the *Liber Hermetis*, chapter 2. These are based on the dwadasamsas of the signs of the first type described above in which there are 12 in each signs starting with the sign itself. The sex of the dwadasamsa determines the sex of the degree or portion thereof.

[1] Both classifications match the list given in Al-Biruni, p. 271, called 'degrees which increase and diminish fortune'.

[2] The constellation of the River Eridanus lies south of the Ram. *S.N.* p. 217: "**Achernar**... the End of the River, nearly its present position in the constellation about 32° from the south pole; but the title was first given to the star now lettered θ, the farthest in the stream known by Arabian astronomers." *S.N.* p. 219: Under the heading "θ, double, 3 and 5.25" "**Archernar** was the early name for this [star] at the then recognized end of the stream. . ." "It is the solitary star visible from the latitude of New York City [40°N 45'] in early winter evenings, low down in the south, on the meridian with Menkar of the Whale; but Baily said that its brilliancy was probably lessened since Ptolemy's time, for the latter designated it by α, i.e., of the first magnitude."

[3] The Bull is depicted without its hind quarters.

[4] Melancholy.

the even one, of the flavors it is sweetness with astringency, of the colors it is green and white, of the animals it is the quadrupeds with cloven hoofs.

Of plants, in its share are the tall trees, and all trees that bear fruit, and all trees that need little water, and every tree that grows on mountains and bears useful fruit, and every planted tree whose taste and fragrance are good.

Of the regions, in its share is the fifth,[1] and the land of Kush,[2] and Mahin, and Hamdhan,[3] and Acart, and Africa from the Qreevan to Asarbalus, and Alkufa, and Al-Batzra, and Assouan in Egypt.

Of the plains it has all cultivated land plowed with cattle and elephants, and the gardens and orchards, and the places of all plants with good fragrance, and all the places that have warm temperate soil where plants grow.

Of the letters it has *dalet* (ד) and *samekh* (ס). Its years are 8 and so is the number of its months, and its days are 20, and its hours are 16.

In the first face there ascends the Hero who has a Sword in his left hand and a Club in his right hand, and two Candles on his shoulders.[4] There also ascends a Great Ship,[5] above it a Lion, and a naked man is sitting there, and below the Ship there is half of the body of a dead woman; there also ascends another human figure. The Hindus say that it is a woman with hair, who has a son, and who wears clothes partly burnt.[6] According to Ptolemy there ascends half of the Bearer of the

---

[1] This is a reference to the fifth clima. See page 15, note 6. [RH]

[2] Possibly Ethiopia.

[3] The names of some of these places were unfamiliar to me, so I transcribed them phonetically, with possible vowel errors. On the whole, Ibn Ezra seems to follow the Arab astrologers scheme, possibly combined with Ptolemy's one for allotting the regions on the Earth to the signs. This scheme basically follows Ptolemy's designation of the Earth triplicity to the south-east quadrant. Ibn Ezra's list seems to fit Al-Biruni's tables, as shown in *The Book of World Horoscopes,* new edition p. 508, 514, and Al-Biruni, p. 220. See bibliography.

[4] These would be two bright fixed stars — Betelgeuze, α Orionis, in the right shoulder of Orion, and Bellatrix, γ Orionis, in the left shoulder.

[5] The constellation Argo.

[6] *Picatrix:* "And there ascends in the first face of Taurus a woman of curly hair, having a single child who is dressed in clothes like unto fire, and she herself dressed in similar clothes. And this is a face of plowing and working the

Devil's Head, the tail of the Ram, and the flow of the water at the end of the River.

In the second face there ascends a Ship, and a naked man carrying a key in his hand. There also ascends the second half of the dead woman's body. Hindu astrologers say that there ascends a man that resembles a ram in his face and in his body, who has a wife that resembles an ox. His fingers are like goat's hoofs, and that man is very hot and gluttonous, and does not give his soul any rest. He builds land and drives the oxen to plow and sow.[1] There also ascends a beautiful figure, with a wand in her right hand and she is raising her left hand.[2] According to Ptolemy there ascends the knee of the Bearer of the Devil's Head, his legs and his feet, and the back of the Bull and its horns, and its hand and its belly and its right foot, and the beginning of the River and its middle.

In the third face there ascends the end of the body which resembles

---

earth, of sciences, geometry, of sewing seed, and making things. And this is its form."

Agrippa: "In the first face of Taurus ascendeth a naked man, an Archer, Harvester or Husbandman, and goeth forth to sow, plough, build, people and divide the earth, according too [sic] the rules of Geometry; . . ." Obviously there is some discrepancy here between the *Picatrix* and Agrippa, but Ibn Ezra's image goes more toward the *Picatrix*. [RH]

[1] *Picatrix:* "And there ascends in the second face of Taurus a man like the figure of a camel and having on his fingers are hooves like those of cows, and he is covered completely with a torn linen sheet. He desires to work the land, to sow, and to make things. And this is a face of nobility, power, and of rewarding the people. And this is its form."

Agrippa: ". . . in the second face ascendeth a naked man, holding in his hand a key; it giveth power, nobility, and dominion over people: . . ." [RH]

[2] This female figure can be found in the description of Bellatrix, the star in the left shoulder of Orion, in *S.N.* p. 313, 314: "**Bellatrix,** the Female Warrior, the Amazon Star, is from the translation, rather freely made in the *Alfonsine Tables*, of its Arabic title, **Al Najīd,** the Conqueror." "In astrology it was the natal star of all destined to great civil or military honors, and rendered all women born under its influence lucky and loquacious; or, as old Thomas Hood said, 'Women born under this constellation shall have mighty tongues.' "

the Head of a Dog,[1] and a man standing with an animal in his hand, and he has two carts, on which is seated a young man, and two horses are pulling the carts, and the young man has a sheep in his hand.[2]

Hindu astrologers say that there ascends a man whose feet are white and so are his teeth, which are so long that they can be seen outside his lips. His complexion is reddish and so is his hair, and his body resembles that of an elephant and a lion, and he is not reasonable, and all his thoughts are [towards] evil, and he is sitting propped up. There also ascends a horse, and a dog, and a small calf.[3] According to Ptolemy there ascends the right foot of the Bearer of the Devil's Head, and the shoulder of the One Holding the Reins, and his head, and his knees, and his left hand, and his horn,[4] and the end of the Bull, and the beginning of the River.

One born in it will have an erect posture, long face, large eyes, thick and short neck, broad forehead, sharp nostrils, and curly hair. There is confusion in his speech and also in his mind. His hair is dark and his limbs are short. He is lustful, and gluttonous, and antagonistic.[5] A person born in one of the degrees of the Pleiades, which are from 13°

---

[1] This could be a reference to Sirius, the Dog star in Canis Major. *S.N.* p. 121: "It has been asserted that Ovid and Vergil referred to Sirius in their **Latrator Anubis**, representing a jackal- or dog-headed Egyptian divinity, guardian of the visible horizon and of the solstice."

[2] This could be Auriga, the Charioteer, described as carrying a goat on the left shoulder and kids on the wrist. *S.N.* p. 83.

[3] *Picatrix:* "And there ascends in the third face of Taurus a man of ruddy coloring with large, white teeth appearing outside of his mouth, and a body like an elephant whose legs are long; and there ascends with him one horse, one dog, and one calf. And this is a face of laziness, poverty, misery, and fear. And this is its form."

Agrippa: ". . . in the third face, ascendeth a man in whose hand is a Serpent, and a dart, and is the image of necessity and profit, and also of misery and slavery." [RH]

[4] Capella — a star in Auriga, the Charioteer. *S.N.* p. 86, 87: "**Capella,** . . . **Capra,** . . . **Amalthea** came from the name of the Cretan goat, the nurse of Jupiter. . ." "Others said that the star represented the goat's horn broken off in play by the infant Jove and transferred to the heavens as **Cornu Copiae**, the Horn of Plenty."

[5] Probably referring to the reputed Taurus stubbornness.

to 15 °,[1] will have eye disease.[2]

One born in the first face will be short, with large eyes and thick lips, and he has a mark on his neck, and another one on his genitals. He is generous, and his friends are many, and he enjoys all kinds of pleasure.

One born in the second face will have round face, wide chest, beautiful eyes, generous character, and intelligence. There is hair on his shoulder and a mark on his waist, and some have pain in the arteries.

One born in the third face will be handsome in body and face, and a mark in his left eye. And he works hard and has no luck with women. One born in the last degrees of the sign will be a castrate or androgynous.

In its share of the human body are the neck and the throat, and all the ailments that occur in these parts, such as mumps, and torticollis, and goiter.[3]

According to Egyptian astrologers the ailment of Saturn in this sign is in the heart, Jupiter's is in the stomach, Mars' is in the neck, the Sun's is in the knees, Venus's is in the head, Mercury's is in the feet, and the Moon's is in the legs. And of the signs of ailments it is like Aries.[4]

In its share among people are the medium [sized], and all those who lust for intercourse, and for eating, drinking, making merry and dancing.

It is the domicile of Venus and the exaltation of the Moon in the 3rd degree. The north node of Mars is in the 10th degree, and it is [Mars'] house of detriment.

The ruler of the triplicity by day is Venus and then the Moon, and by night it is the Moon and then Venus, and their co-ruler by day and by night is Mars.

The first face according to the Egyptian and Persian astrologers belongs to Mercury, the second to the Moon, and the third to Saturn. According to the Hindu astrologers the first face belongs to Venus, the

---

[1] Now in the last degrees of Taurus. [RH]

[2] That area has a nebula, which traditionally causes blindness or eye problems.

[3] The Latin has "scrofula, swelling of the throat, and the contractions and wrinkling of these." [RH]

[4] This attribution is unfamiliar to me.

second to Mercury, and the third to Saturn.[1]

These are the bounds according to the Egyptian astrologers and all [other] astrologers: To Venus is [allotted] up to [the first] 8 degrees, then to Mercury 6, to Jupiter 8, to Saturn 5, and to Mars 3 degrees. According to Ptolemy to Venus 8 degrees, to Mercury 7, to Jupiter 7, to Mars 2, and to Saturn 6 degrees.

The rulership of the 1st and the 2nd one-ninth [of the sign belongs] to Saturn, the 3rd to Jupiter, the 4th to Mars, the 5th to Venus, the 6th to Mercury, the 7th to the Moon, the 8th to the Sun, and the 9th to Mercury.[2]

The rulership of the 1st one-twelfth [of the sign belongs] to Venus, the 2nd to Mercury, the 3rd to the Moon, the 4th to the Sun, the 5th to Mercury, the 6th to Venus, the 7th to Mars, the 8th to Jupiter, the 9th and the 10th to Saturn, the 11th to Jupiter, and the 12th to Mars.

From the beginning of the sign until the end there are 3 mixed degrees, then 2 bright ones, then 2 empty ones, then 8 bright ones, then 5 empty ones, then 3 bright ones, then two mixed ones.[3]

From the beginning of the sign until 7 degrees [they are] male, then 8 female, then 15 male.

The pits of the stars are the 5th, 12th,[4] 18th, 24th, 25th, and 26th degrees. The degree that bestows grace and honor is the 3rd, the 15th, 27th, and the 30th.

Of the superior stars in this sign there are: The Back of the Leper[5] at 14° and latitude of 22° 20' north, of the second magnitude and of the nature of Venus. The Devil's Head.[6] The one at 15° and latitude of 23° 3' north is of the second magnitude, and its nature is the co-mixture of Jupiter and Saturn. There is also the bright star called the Left Eye of

---

[1] This time the text has the two systems of decanic rulers attributed correctly. See page 23, note 1. [RH]

[2] A reminder: The rulership scheme of Navamsas is to begin with the ruler of the first cardinal sign in the triplicity of the sign being divided.

[3] These do not add up to 30. [Additional by RH] The Latin text differs considerably and is probably less reliable. But for the record here they are. "From the beginning of the sign there are 3 mixed or middling degrees, 6 bright degrees, 2 void, 8 bright, 5 void, 4 bright, and 2 mixed."

[4] The Latin has the 13th degree rather than the 12th. [RH]

[5] I was unable to identify this star.

[6] Algol.

the Ox[1] at 28° and latitude of $5^1/_2$° south, of the first magnitude, and its nature is the co-mixture of Mars and Venus. It is [one] of the deadly stars when the significator of life reaches it.

Of the dark stars it has one at 10° and latitude of 20° 40' north, and another star, just as dark, at 12° and latitude of $40^1/_{12}$° north, [also] of the deadly ones.

Gemini is of the air element, masculine, of the diurnal signs, occidental, of the hot season, and it has two bodies.[2] Its [temporal] hours grow longer than the even ones, and at its end the day is the longest in the year. In all regions it has short ascension. It indicates an upright nature [that promotes] growth, perfect abundance, and anything that's useful to animals and plants.

On the whole its nature is temperate. Its front part is humid, its middle is temperate, and its end is variable. In the north side it produces earthquakes and winds,[3] and in the south it produces pestilent [weather]. It is of human form, with perfect limbs,[4] and it is barren.

In its share is the right side of the west and the westerly wind, and its nature is warm and moist.[5]

[Of the body] in its share is the blood, of the flavors it is the very sweet, of the colors it is those that change into many other colors, of the living things it is man, and monkey, and the bird that has a pleasant song, and of the plants in its share are the tall trees. In general it indicates everything that is high, like the sky, and the air, and the wind.

Of the regions in its share is the sixth one,[6] and the land of Iran, and Armenia, and greater Abidian, and Yalan, and Brian, and Barqa, and Khalon, and the Land of Israel, and Egypt, and Ispahan, and Kirman.

---

[1] Aldebaran.

[2] That is, it is bicorporeal. [RH]

[3] Castor and Pollux in the constellation of the Twins are associated with stormy weather and ship wrecks, and are considered the patrons of seafarers. Mercury, its ruler, is also a wind producing planet.

[4] When the image in the constellation is depicted with missing parts of the body, it is reflected in its astrological signification as we have seen in the case of Aries and Taurus before.

[5] The Air element is a combination of heat and moisture.

[6] The sixth clima. See page 15, note 6. [RH]

In its share are all steep and tall mountains, and the places of bird hunters, and [places of] dice players, and of musicians.

Its letters are *gimel* (ג) and *a'yin* (ע). Its years are 20 and so are its months, the days are 50, and the hours are 4.

In the first face there ascends the tail of the figure whose head resembles a dog's head, and a man with a wand in his hand. In the south there ascend two carts drawn by two horses on which a man is sitting and driving them.[1] There also ascends the head of the Unicorn. Hindu astrologers say that there ascends a beautiful woman standing in the air, and she can sew.[2] There ascends according to Ptolemy the head of the Rein Holder and his right foot, and the Bull's horn, and the shoulder of the Dog and its left foot, and the head of the Hare and its paws.

In the second face there ascends a man who has a golden musical instrument which he is playing. There ascends an animal standing on a tree, and a wolf that has a mark on its paw. Hindu astrologers say that there ascends a black man with his head bound in lead, and a weapon in his hand, and an iron helmet on his head, and on the helmet there is a silk crown, and in his hand a bow and arrows. He likes ridicule and mockery, and he walks around in a garden that has trees and flowers, and in his hand scales stones. He strikes them with his hand, and plays [music], and picks flowers from the garden.[3] According to Ptolemy

---

[1] Castor, one of the Twins' stars, was known as the Horseman, but this could also be some mix up with Auriga, the Charioteer, who is actually north west of the Twins, and can be seen from northern latitudes.

[2] *Picatrix:* "And there ascends in the first face of Gemini a beautiful woman, a mistress of sewing; and with her ascends two calves and two horses. And this is a face of the art of the scribe, of reckoning, of number, of giving and receiving [i.e., trade], and of the sciences. And this is its form."

Agrippa: "In the first face of Gemini ascendeth a man in whose hand is a rod, and he is, as it were, serving another; it granteth wisdom, and the knowledge of numbers and arts in which there is no profit: . . ." Agrippa's images are beginning to deviate from Ibn Ezra and the *Picatrix.* [RH]

[3] *Picatrix:* "And there ascends in the second face of Gemini a man whose face is like an eagle and his head is covered with a linen cloth; he is garbed and protected in a leaden cuirass, and on his head an iron helmet upon which is a silken wreathe; and [he is] holding in his hand a bow and arrows. And this is a face of burden and also of evils and of subtlety. And this is its form."

Agrippa: ". . . in the second face ascendeth a man in whose hand is a Pipe,

there ascends the right hand of the Rein Holder, the legs of the later[1] Bull, and the hand of the Giant, and his shoulder, and his head, and his chest, and his girdle, and his knees, and hind leg, and the belly of the Hare and its tail.

In the third face there ascends the figure of a frightened man with a turban on his head, and in his hand a musical instrument with strings of gold. There also ascends a barking Dog,[2] and a fish called dolphin, and an image of a monkey, and the thread of a tailor, and the first half of the Small Bear, and the tail of the Unicorn that is wrapped in the root of the Virgin.[3] Hindu astrologers say that there ascends a man seeking arms who has a bow and a quiver, and in his hand an arrow and clothing and golden ornaments, and he desires to play music and laugh and mock in all sorts of ways.[4] According to Ptolemy there ascends the shoulder of the later Twin and his hand and his posterior part and his right foot and his genitalia, and the tail of the Hare, and the mouth of the Dog and its paws and its right legs, and the first oar of the Ship and some of the second oar.

---

and another being bowed down, digging the earth, and they signify infamous and dishonest agility, as that of jesters and jugglers; it also signifies labours and painful searchings: . . ." [RH]

[1] We do not know of a second bull in that area, but in *S.N.* p. 378, 379, regarding Taurus, we find: ". . . the Bull generally being drawn with only its forward parts, . . ." "This incomplete figuring of Taurus induced the frequent designation in early catalogues, **Sectio Tauri**, which the Arabians adopted, dividing the figure at the star *o*, but retaining the hind quarters as a sub-constellation."

[2] Possibly Canis Minor, the Lesser Dog, in some descriptions announcing the rising of the Canis Major.

[3] The Unicorn is quite far from the Virgin, but the tip of its horn is not too far from the head of Hydra, the Water Snake, whose body is extremely long and reaches all the way to the feet of the Virgin.

[4] *Picatrix:* "And there ascends in the third face of Gemini a man garbed with a cuirass holding a bow and arrows, and a quiver. And this is a face of boldness, honesty, and the division and alleviations of labor. And this is its form."

Agrippa: "In the third, ascendeth a man seeking for Arms, and a fool holding in the right hand a Bird, and in the left a pipe, and they are the significations of forgetfulness, wrath, boldness, jests, scurrilities, and unprofitable words: . . ." [RH]

One born in this [sign] will be erect in stature [with] a wide chest and handsome body and face. He has eloquent speech and a strong voice, a generous character, broad shoulders, beautiful eyes, and curly hair. He is fast in his work and skillful in all crafts. Among them there are writers and mathematicians, astronomers and astrologers, and great scientists. Such a person is trustful and moderate in his fear of God.[1]

One born in the first face will have a handsome body and eyes and hair, with a mark on his head or cheek, and his countenance will be sharp, and he does not get angry, and he works hard, and has no luck with women.

One born in the second face will be short in stature, and dark [looking], with a black mark under his arm. His speech is eloquent and he is a man of morals, handsome, generous, and associates with kings

One born in the third face will have narrow face and small eyes. He is boastful, lascivious, says improper things, and is a liar. One born at the end of the sign will not have perfect eyes.[2]

In its share of the human body are the shoulder blades, the arms, the hands, and the shoulders. Its ailments are all those of the blood and those that occur in the aforementioned organs. According to the Egyptian astrologers the ailment of Saturn [in it] is the stomach, Jupiter's is in the hips, Mars' is in the shoulder, the Sun's is in the thighs, Venus's is in the neck, Mercury's is in the head, and the Moon's is in the feet.

In its share of people are the kings, the magnates, the heroes, the magicians, the sorcerers, and those with the icons and amulets. [Also in its share are] all things of play and music, and hand craft and all delicate skills.

It is the domicile of Mercury and the exaltation of the North Node in the 3rd degree, and the detriment of Jupiter, and the high position of the Sun, and the place of Venus at 27° at the present time, and the inferior position of Saturn at 12°.[3]

The ruler of the triplicity by day is Saturn then Mercury, and by

---

[1] Not very religious.

[2] Maybe because of the presence of the Beehive nebula in the Crab in that area. But this is not certain since the latter part of tropical Gemini was still in the constellation of the Twins at this time.

[3] These are values of the auges or oppositions of the auges. See page 22, note 1.

night it is Mercury then Saturn, and their co-ruler by day and by night is Jupiter.

The first face according to the Egyptians and the Babylonians belongs to Jupiter, the second to Mars, and the third to the Sun. According to the Hindu people the first belongs to Mercury, the second to Venus, and the third to Saturn.

These are the bounds according to the Egyptians and [other] astrologers: to Mercury [are allotted] 6°, Jupiter 6°, Venus 5°, Mars 7°, and Saturn 6°. According to Ptolemy Mercury has 7°, Jupiter 6°, Venus 7°, Mars 6°, and Saturn 4°.

The 1st one-ninth belongs to Venus, the 2nd to Mars, the 3rd to Jupiter, the 4th and 5th to Saturn, the 6th to Jupiter, the 7th to Mars, the 8th to Venus, and the 9th to Mercury.

The influence of the 1st one-twelfth belongs to Mercury, the 2nd to the Moon, the 3rd to the Sun, the 4th to Mercury, the 5th to Venus, the 6th to Mars, the 7th to Jupiter, the 8th and 9th to Saturn, the 10th to Jupiter, the 11th to Mars, and the 12th to Venus.

From the beginning of the sign to the end there are 7 bright degrees, then 3 mixed ones, 7 empty ones, then 6 bright ones, and 6 mixed degrees.[1]

From the beginning of the sign to the end there are 6 female degrees, then 11 male, then 6 female, then 4 male, and 3 female.

The pitted degrees are the 2nd, 12th, 17th, 26th, and the 30th. The degree that bestows grace and honor is the 11th.

There is a star in it on the left foot of the Twins at 5°, with latitude of 31° 40' south, of the 1st magnitude, [similar in nature to] the co-mixture of Mars and Venus.[2] Another star is on his left shoulder, at 5°, with latitude of $17^1/_2$° south, of the 2nd magnitude, [similar in nature to] the co-mixture of Mars and Mercury.[3] There is the star of the Rein

---

[1] The preceding list of degrees do not add up to 30. The Latin text has "7 degrees are bright, 3 mixed, 5 bright, 3 void, 6 bright, and 6 mixed." [RH]

[2] The Twins are a northern constellation, so I was unable to determine whether this is a scribal error or reference to another star, maybe co-rising with the foot of the Twin. This confusion appears in the stars mentioned below too.

[3] Here, again, the latitude does not fit the Twins, but the rest of the description matches that in the *Tetrabiblos,* Book I, chapter IX, Ashmand trans., p. 25: "The stars in the feet of Gemini have an influence similar to that of Mercury, and moderately that of Venus. The bright stars in the thighs are like

Holder at the 10th degree, with latitude of 16° north, of the 1st magnitude, [similar in nature to] the co-mixture of Jupiter and Saturn. There is the great star called *Al-Ayuk*, at the 11th degree, with latitude is 22½° north, of the 1st magnitude, [similar in nature to] the co-mixture of Jupiter and Saturn. There is the middle star in the girdle at the 15th degree, with latitude of 24° 45' south, of the 2nd magnitude, [similar in nature to] the co-mixture of Jupiter and Saturn. There is the shoulder of the right Twin at the 16th degree, with latitude of 17° south,[1] of the 1st magnitude, [similar in nature to] the co-mixture of Mars and Mercury. There is the foot of the right Twin at 24°, with latitude 11½° south, of the 1st magnitude, [similar in nature to] the co-ixture of Mars and Venus. And there is the Dog called Al-Shari Al-Abur,[2] and of the dark stars there is the Cloudy one on the head of the Giant at 12°, with latitude of 13° 50' south,[3] and it is of the deadly ones.

Cancer is of the water signs, female, of the nocturnal signs. It is northern, of the summer signs, and changeable because time changes in it. At its beginning the day begins to decrease and the night to increase. Its [temporal] hours are longer than the even ones. It has straight ascension which increases.[4]

On the whole it indicates moderate cold and moisture[5] [suitable] for growth and multiplication, and it has some degrees that are moist and hot. Its front part is dry and destructive, its middle is temperate, and its end is moist. When it is northern (the northern part), it heats and burns, and when it is southern, it is moist.

Saturn. Of the two bright stars on the heads, the one which precedes and is called Apollo (Castor) is like Mercury; the other follows, called Hercules (Pollux), is like Mars."

[1] Should be 'north', as the Twins' shoulder is north of both equator and ecliptic.

[2] The text has *Al-Shari Al-Abud* (אלשערי אלעבוד), but the letter *resh* (ר) can easily get corrupted into *dalet* (ד). Based on *S.N.* p. 119, 120-121, this is probably Sirius, the Dog star in Canis Major.

[3] If the Giant is Orion, then it would be south of the ecliptic, since Orion straddles the equator with his head in the north.

[4] That is, the degrees later in the sign have a longer ascension than the ones nearer the beginning. [RH]

[5] The water element is cold and moist.

It is of the aquatic animals' form, and in its share are the center of the northern [direction] and the northern wind. Its nature is cold and moist, and it indicates the phlegm in people and everything of cold and moist nature, somewhat moderate. Of the flavors [it indicates] the sour and the salty,[1] of the colors [it indicates] the white and the color of soil and the smoke-like colors and those similar to them. In its share of living things are all the aquatic animals and the shrimp and the small fish, and [of] the land animals the scorpions and [other] small crawling creatures. In general it indicates water that has a lot of movement, and all vegetation that is near water, and rain water, and all fresh water.

Of the regions[2] it has the seventh, and the land of Armenia Minor which is beyond Muqan and Adrianan, and eastern Kourassa, and Tzin, and parts of Africa, and Balkh.

In its share are the lakes and all sea-shores and all river banks, and all medium height trees.

Of the letters it has *dalet* (ד) and *peh* (פ). Its years are 25 and so are its months, and its days are 5 and so are the hours.

In the first face there ascends the rear half of the Great Bear, and a whole figure wrapped in clothes, near the [previously] mentioned [figure]. There ascends an iron pig with a head of copper, and a maiden. Hindu astrologers also said that there ascends a handsome young man, wearing clothes, and he has [some] sickness, and in his face and fingers there is some crookedness, and his body resembles that of a horse and an elephant, and his feet are white, and on his body are hanging ornaments in the shape of trees, and he sits in an orchard that grows fragrant stalk.[3] According to Ptolemy there ascends the face of the Great Bear, and the head of the first and the second Twins, and the posterior part of the first Twin and his hands, and the Small Dog, and the

---

[1] In most other sources all of the water signs are associated with only the salty taste. [RH]

[2] The seventh clima. See page 15, note 6. [RH]

[3] *Picatrix:* "And there ascends in the first face of Cancer a man having twisted and crooked fingers and head; and his body is like the body of a horse, and having white feet and upon his body fig leaves. And this is a face of teaching, knowledge, of love, subtlety and of skills. And this is its form."

Agrippa: ". . . In the first face of Cancer ascendeth the form of a young Virgin, adorned with fine cloathes, and having a Crown on her head; it giveth acuteness of senses, subtlety of wit, and the love of men: . . ." [RH]

remainder of the Great Dog and the belly of the Ship.

In the second face there ascends a maiden that resembles a Cloud.[1] There also ascends half of the Dog and half of the ears of the northern Ass.[2] Hindu astrologers say that there ascends a beautiful maiden with pleasant speech, on her head a myrtle crown, in her hand a wooden stick, and she desires wine and music.[3] According to Ptolemy there ascends the head of the Great Bear, the rear side of the Crab, and the belly of the Ship.

In the third face there ascends a maiden sometimes going east and sometimes going west. There also ascends the later Dog, and the second half of the ears of the northern Ass, and the second, southern Ass. Hindu astrologers say that there ascends a man whose foot resembles that of an animal, and on his body there is an animal, and he intends to enter a ship to go to sea and bring gold and silver to make rings for his wives.[4] [[According to Ptolemy]] there ascends the neck of the Great Bear and its right paw, the horns of the Crab, and the head of the Fighting Hero, and the end of the Ship.

One born under this [sign], his limbs are thick, his forehead is

---

[1] This could be the nebula in the Crab, known as the Beehive, Praesaepe, which was also called the Cloudy One, Little Cloud, Little Mist. *S.N.* pp. 112, 113. I was unable to find the 'maiden'.

[2] Two fixed stars, γ and δ in the constellation of the Crab, are called the Aselli, the Two Asses or the Donkeys. *S.N.* pp. 110, 111.

[3] *Picatrix:* "And there ascends in the second face of Cancer a woman of beautiful visage, and having on her head a green wreath of myrtle, and in her hand is the stem of the planet which is called the water lily, she is singing songs of love and joy. And this is a face of playing, and of wealth, joy and abundance. And this is its form."

Agrippa: ". . . in the second face, ascendeth a man cloathed in comely apparel, or a man and woman sitting at the table and playing; it bestoweth riches, mirth, gladness, and the love of women: . . ." [RH]

[4] *Picatrix:* "And there ascends in the third face of Cancer Celhafe [?] and he holds a serpent in his hand holding before him golden chains. And this is a face of running, riding, and acquiring in war in strife and contrariety. And this is its form."

Agrippa: "in the third face ascendeth a man a Hunter with his lance and horne, bringing out dogs for to hunt; the signification of this is the contention of men, the pursuing of those who fly, the hunting, and possessing of things by arms and bravelings."

large, his teeth are far apart, and he is mute and deaf. He likes people and is respectable. In the nativity of women this is not good for it indicates difficult things.[1] One born in the first face will be handsome in body and hair. His eyebrows are stuck together, his nostrils are long, his shoulders are broad, there is a mark under his elbow or in his right arm. He has a kind nature, and his friends are many, and he is cunning. One born in the second face will be red, short in stature, beardless, with a black mark in (under) his eyes, and he likes people. One born in the third face will be fat and short, with much hair in his eyebrows, wide chest and big belly. He will be strong, and at times will be seized by a heart pain, and he will exhaust himself very much. One born at the end of the sign will not have good luck.

Its share in the human body are the chest, and the breasts, and the upper bowels (stomach), and the ribs, and the spleen, and the lung, and of the ailments, everything that afflicts these organs. It also has heaviness in the eyes (impaired vision), and at 22° there is a Cloudy star that denotes eye ailment and defect. The whole sign indicates pruritus,[2] itch, leprosy, pock-marks, baldness and thin beard. Of People, in its share are all inferior persons, and common people, and sailors, and travelers on the road. According to Enoch it is the sign of the world.[3]

According to Egyptian astrologers Saturn's ailment [when found in it] is in the hips, Jupiter's is in the genitalia, Mars' is in the upper bowels (stomach), the Sun's is in the feet, Venus's is in the hands, Mercury's is in the neck and the Moon's is in the head.

It is the domicile of the Moon, the exaltation of Jupiter at 15°, the fall of Mars at 28°, and the detriment of Saturn, and [the place of] the north node of Jupiter at 9°, and the north node of Saturn at 19°. The

---

[1] This is an intriguing comment without further explanation.

[2] An itching disease without obvious marks or rash. [RH]

[3] *S.N.* p. 107, regarding the constellation of Cancer: "It is the most inconspicuous in the zodiac. . . Yet, few heavenly signs have been subjects of more attention in early days, and few better determined; for, according to Chaldean and Platonist philosophy it was the supposed **Gate of Men** through which souls descended from heaven into human bodies." "In astrology . . . it has been the **House of the Moon**, from the early belief that this luminary was located here at creation; and the Horoscope of the World, as being, of all signs, nearest to the zenith." [Additional by RH] The idea of Cancer being the world Ascendant (what 'horoscope' correctly means) is found throughout Greek language astrology and is always attributed to the Egyptians.

rulers of the triplicity by day are Venus followed by Mars, and by night Mars followed by Venus, and their co-ruler by day and night is the Moon.

The first face according to the Egyptians and the Babylonians [belongs] to Venus, the second to Mercury and the third to the Moon. According to Hindu astrologers the first [face belongs] to the Moon, the second to Mars, and the third to Jupiter.

These are the bounds according to the Egyptians and [other] astrologers: to Mars 6°, to Jupiter 6°, to Mercury and Venus 7°, and to Saturn 4°.[1] According to Ptolemy to Mars 6°, to Jupiter 6°, to Mercury and Venus 7°, and to Saturn 4°.[2]

The [influence of the] 1st one-ninth [belongs] to the Moon, the 2nd to the Sun, the 3rd to Mercury, the 4th to Venus, the 5th to Mars, the 6th to Jupiter, the 7th and 8th to Saturn, and the 9th to Jupiter. The influence of the 1st one-twelfth [belongs] to the Moon, the 2nd to the Sun, the 3rd to Mercury, the 4th to Venus, the 5th to Mars, the 6th to Jupiter, the 7th and 8th to Saturn, the 9th to Jupiter, the 10th to Mars, the 11th to Venus, and the 12th to Mercury.

From the beginning of the sign until the end there are 7 mixed degrees, then 5 bright ones, then 2 mixed ones, then 4 bright ones, then 2 dark ones, then 8 bright ones, and then 2 dark ones.[3]

From the beginning of the sign until the end there are 2 male degrees, then 2 female ones, then 11 male ones, then 4 female ones, then 3 male ones.[4]

The pitted degrees are the 12th, 17th, 23rd, 26th, and 30th. The degrees that bestow grace and honor are the 1st, 2nd, 3rd, 14th, and 15th.

---

[1] These do not add up to 30°. The error is in the Hebrew. But the Latin also has an error: ". . . of Mars 7°, of Venus 6°, of Mercury 6°, of Jupiter 7, and of Saturn 8°." The Hebrew should have had 7° for Mars, but the Latin should have 4° for Saturn at the end. [RH]

[2] The Latin incorrectly gives Saturn 3°. [RH]

[3] The Latin incorrectly omits the 8 bright and 2 dark degrees at the end. [RH]

[4] This time the Latin appears to be correct. "And from the beginning of the sign up to the end 2 degrees are masculine, 6 feminine, 2 masculine, 2 feminine, 11 masculine, 4 feminine, and there are 3 masculine," The Hebrew appears to have omitted the first 6 feminine and 2 masculine degrees. The Latin adds up to 30°. [RH]

Of the superior stars it has the Dog called *Al-Sha'ari Al-Abur* at 3° at the present time, with latitude 39° 10' south, of the 1st magnitude, [similar in nature to] the co-mixture of Mars and Jupiter. There is also the head of the early Twin, at 8° at the present time, with latitude of 9° 40' north, of the 2nd magnitude, [similar in nature to] the co-mixture of Jupiter and Mercury. There is the head of the later Twin, at 12° at the present time, with latitude of 6° 15' north, of the 2nd magnitude, of the nature of Mars alone. There is the Dog called *Al-Sha'ari Al-Gumeiza,*[1] at 15° at the present time, with latitude of 16° 10' south, of the 1st magnitude, [similar in nature to] the co-mixture of Mars and Mercury. [Also found] there are 4 of the dim stars which are close to the [[Great]] Bear. One is at 12° at the present time, with latitude of 22° 50' north; the second one is at 17° at the present time, with latitude of 22° 2' north; the third one is at 27° at the present time, with latitude of 22° 45' north, and the fourth is at the end of the sign, with latitude of 20° north. There is a star on its belly (the Crab's) at 26°, with latitude of 6° north.

Leo is of the fire signs, male, of the diurnal signs, oriental, of the summer signs, fixed (for time is stationary in it), and its [temporal] hours are longer than the even ones. Its steps (ascension) are even and its steps are long.

Its nature is burning and destructive heat. Its beginning is a little warm, its middle is very destructive and producing sickness, and its end produces winds. When it is northern, it is scorching hot, and when it is southern, it is moist.

It is of the four-legged forms that have hoofs and prey, and its limbs are severed, and it has half a voice. In its share is the left of the east, and the easterly wind. Its nature is hot and dry, and [in the body it indicates] the red bile. Its flavor is the bitter and sharp, and of the colors it is the saffron and yellow. Of the animals, [in its share] are the lions, the tigers, the hyenas, the bears, and the wolves. Of the metals it is gold and silver, and precious stones, and the stone that pierces [precious] stones (diamonds) called *Al-mas,* and the stone that gathers straw (magnet) called *Al-budai,* and every fiery occupation.

Of the regions in its share is the fourth,[2] and Baghdad, and Persia,

---

[1] β Canis Minor. *S.N.* p. 134.

[2] The fourth clima. See page 15, note 6. [RH]

and Turkey as far as the inhabited places, and Nisbur, and Tarsus, and every place that is difficult to climb, and every rugged place, and palaces of kings, and strong fortresses, and every tall and steep mountain, and every place of open air.

Of the letters [it has] the *heh* (ה) and the *kof* (ק) Its years are 19 and so are its months. The days are $47^1/_2$ and the hours are 23.

In the first face there ascends a Bear, and a Dog with a bow on its back, and half a Ship and her sailors, and the head of a black Animal, and the head of a Horse[1] and the head of an Ass. Hindu astrologers say that there ascends a large tree on whose branches there is [[a dog and]] the bird called vulture, and a man wearing pretty clothes, though dirty, and he is about to hit his father. There also ascends the master of the horse that looks to the north.[2] According to Ptolemy there ascends the neck of the Great Bear [[and its left hand]], the head of the Lion, the neck of the Hero and half of the Ship.

In the second face there ascends a figure with his hands raised up and he is shouting with a loud voice, playing music, and dancing. There also ascend two vessels of wine, and a glass goblet, and a musical instrument made of deer's horns, and the second half of the Ship, and the eye of the Animal, and the middle of the Horse, and the middle of the Ass. Hindu astrologers say that there ascends a man whose nostrils are fine, on his head there is a shape of a crown of white myrtle, and a bow in his hand. He is fierce like a lion in his anger, and he is wrapped in a cloak that looks like a lion.[3] According to Ptolemy there

---

[1] *S.N.* p. 247: The Animal is another name for Hydra, the Water Snake. As for the Horse, *S.N.* p. 249: "Al Sufi mentioned an early Arab figure, **Al Ha'il**, the Horse, formed from stars some of which now belong to our Hydra, but more to Leo and Sextans".

[2] *Picatrix:* "And there ascends in the first face of Leo a man dressed in filthy garments; and there ascends with him the figure of a lord of the horse looking toward the north; and his figure is like the figure of a bear and the figure of a dog. And this is a face of strength, liberality, and victory. And this is its form."

Agrippa: "In the first face of Leo. ascendeth a man riding on a Lion; it signifieth boldness, violence, cruelty, wickedness, lust and labours to be sustained." [RH]

[3] *Picatrix:* "And there ascends in the second face of Leo a man having a crown of white myrtle on his head and in his hand a bow. And this is a face of beauty, of riding, and of the rising up of a man who is ignorant and vile; and

ascends the shoulder of the Great Bear and its right foot, and the neck of the Lion, and the middle of the Hero, and the bow of the Ship.

In the third face there ascends the figure of a Young Man whose occupation is to Drive the Animals, a whip in his hand, and he is driving a Wagon[1] in which a man is sitting, and a small boy is with him, and a garment in his left hand. There ascends a Raven, and the middle of the black Animal, and the end of the Horse, and the end of the Ass. Hindu astrologers say that there ascends an ugly black man, laborious, intelligent, with delicacies in his mouth and meat in his hand.[2] According to Ptolemy there ascends the figure of the Great Bear, and the middle of the Lion, and some of the Fighting Hero.

One born under this sign will have a handsome body; he will be yellow (blonde?) and his eyes are like cat's eyes. He is valiant and irascible, and has a keen appearance, and his legs are [thin]. He is a man of morals and cunning, loves sexual intercourse; he is generous, barren, keeps his word, suffers a lot, puts himself in danger, and he is stubborn. His nature is like that of hyenas; he eats a lot and craves

---

it is a face of war, and of naked swords. And this is its form."

Agrippa: "In the second ascendeth an image with hands lifted up, and a man on whose head is a Crown; he hath the appearance of an angry man, and one that threateneth, having in his right hand a Sword drawn out of the scabbard, & in his left a buckler; it hath signification upon hidden contentions, and unknown victories, & upon base men, and upon the occasions of quarrels and battels: . . ." [RH]

[1] The Great Bear, Ursa Major, is also known as the Wagon (*S.N.* p. 426-429). The young man driving it seems to be in the constellation of Bootes, east of the Wagon. *S.N.* pp. 92-93 mentions other names for Bootes — the Wagoner, the Driver of the Wain, the Herdsman. The following seems to fit the previous image of the man shouting. *S.N.* p. 93: "Others, and perhaps more correctly, thought the word Βοητής, Clamorous, transcribed as **Boetes,** from the shouts of the Driver to his Oxen, — the Triones, — or of the Hunter in pursuit of the Bear."

[2] *Picatrix:* "And there ascends in the third face of Leo an old man black and foul holding fruit and meat in his mouth and a jug covered with copper in his hand. And this is a face of love and delight, and of trays [of food?] and of good fortune. And this is its form."

Agrippa: ". . . in the third face ascendeth a young man in whose hand is a Whip, and a man very sad, and of ill an aspect; they signify love and society, and the loss of ones right for avoiding strife." [RH]

every kind of food. [In] the nativity of women it indicates modesty.

One born in the first face will be handsome in face and body. His complexion is reddish, his eyes are medium, his chest is straight, and so are his legs. He has pains in his upper bowels, and he is known among people, and modest, and associates with kings.

One born in the second face, his body will be handsome, his chest wide and his testicles and thighs thin. He will be bald, and have seizures of pain in his arteries. [He] will be respected among his people and arrogant.

One born in the third face will be somewhat short in stature, with some whiteness and reddishness. His voice is strong, he loves women, has many friends and enemies, and many ailments. One born at the end of the sign will be ugly and full of defects.

In its share of the human body are the chest, the heart, the upper bowels, the arteries, the back, the waist, the ribs, and the nape. According to Egyptian astrologers the ailment of Saturn [in this sign] is in the genitalia, Jupiter's is in the testicles[1] Mars' is in the heart, the Sun's is in the head, Venus's is in the chest, Mercury's is in the arms and the Moon's is in the neck. It also indicates the eyesight and the lower part of the upper bowels, and its diseases are those that occur in the aforementioned organs.

Of people in its share are the kings, the ministers and the nobles, and craftsmen in gold, silver precious stones, and every distinguished skill.

It is the domicile of the Sun, and the detriment of Saturn, and has no exaltation, nor fall of any [other] planet. It is the place of Mars at 12° at the present time.[2] The ruler of the triplicity by day is the Sun followed by the Jupiter, and by night it is Jupiter followed by the Sun, and their co-ruler by day and by night is Saturn.

The first face according to the Egyptians and the Babylonians [belongs] to Saturn, the second to Jupiter, and the third to Mars. According to Hindu astrologers the first [belongs] to the Sun, the second to Jupiter, and the third to Mars.

These are the bounds according to the Egyptians and [other]

---

[1] This might be a scribal error. According to the rulership sequence it should be "the hips". [Additional by RH] The Latin has "in the buttocks."

[2] These are values of the auges or oppositions of the auges. See page 22, note 1.

astrologers: to Jupiter [are allotted] 6°, to Venus 5°, to Mercury 6°, to Saturn 7°, and to Mars 6°. According to Ptolemy to Saturn [are allotted] 6°, to Mercury 7°, to Venus 6°, to Mars 6°, and to Jupiter 5°.[1]

The 1st one-ninth [belongs] to Mars, the 2nd to Venus, the 3rd to Mercury, the 4th to Moon, the 5th to the Sun, the 6th to Mercury, the 7th to Venus, the 8th to Mars, and the 9th to Jupiter.

The influence of the 1st one-twelfth [belongs] to the Sun, the 2nd to Mercury, the 3rd to Venus, the 4th to Mars, the 5th to Jupiter, the 6th and the 7th to Saturn, the 8th to Jupiter, the 9th to Mars, the 10th to Venus, the 11th to Mercury, and the 12th to the Moon.

From the beginning of the sign until its end there are 7 bright degrees, then 3 mixed ones, then 6 dark ones, then 5 empty ones, then 9 bright ones.

From the beginning of the sign until its end there are 5 male degrees,[2] then 2 female ones, then 6 male ones, then 10 female ones, then 7 male ones.

The pits of the stars are the 6th, 13th, 15th, 22nd, 23rd,[3] and the 28th degrees. The degrees that bestow grace and honor are 2,[4] 5, 7, and 17.

Of the superior stars there is the neck of the Warrior at 17° at the present time, with latitude of 20½° south, of the 2nd magnitude, [similar in nature to] the co-mixture of Saturn and Venus. There is the Heart of the Lion at 28° at the present time,[5] with latitude of 10' north, of the 1st magnitude, [similar in nature] to the co-mixture of Mars and Jupiter, and it is of the deadly stars. There is the star called the Back of the Lion at the end of the sign at the present time, with latitude of 13° 40' north, of the 2nd magnitude, [similar in nature] to the co-mixture of Saturn and Venus.

---

[1] In the *Tetrabiblos,* p. 53, the sequence of Mars and Jupiter is reversed, keeping the degree numbers. [Additional by RH] Also the last bound is always supposed to be a that of a malefic.

[2] The Latin has 9 masculine degrees which is clearly an error. [RH]

[3] The Latin omits the 23rd degree. [RH]

[4] The Latin omits the 2nd degree. [RH]

[5] "Present time" ?! The Heart of the Lion, Regulus reached 28° Leo around the turn of the 19th century. [Additional by RH] The Latin has 18° Leo which is about right. Obviously there was a scribal error in which 18 became 28.

Virgo is of the earth signs, female, of the nocturnal signs, southern, of the summer signs, and it has two bodies. Its [temporal] hours are longer than the even ones, and at its end night equals day in all regions. It is straight in its steps, and its steps (ascension) are long. Its nature signifies destruction due to its excessive dryness. Its front part has moisture and is thunderous, its middle is temperate, and its end is dry. When it is northern, it produces winds, and when southern, it is temperate. It is of human form, and fowl form, and it has a strong voice. In its share is the right [side] of the south and the southerly wind. Its (elemental) nature is cold and dry and it produces the black bile.[1] Of the flavors its is the bitter and the astringent, of the colors it is white and purple and earth color. Of living things in its share are humans and fowl, of vegetation, every small plant that has no stalk, like wheat, barley, beans, and other seeds.

Of the regions it has the second one,[2] of the lands it has Aljeremica, and the Euphrates river, and Bosphorus, and Greece. Of the places [it has] every seeded land, the houses of women, and the gardens.

Its letters are *vav* (ו) and *zayin* (ז), its years are 20 and so are its months, its days are 7, and the hours are 4.

In the first face there ascends a beautiful Maiden, her hair is long and in her hand two stalks of wheat. She is sitting on a chair, holding a small child nursing and feeding him.[3] There also ascends a man sitting on the same chair. There also ascends a star called *Shibbolet*[4] behind the

---

[1] In Hebrew 'black bile' is the word for depression. All the earth signs are said to be prone to melancholy.

[2] The second clima but see page 15, note 6. [RH]

[3] *S.N.* p. 462-463, regarding Virgo: "Erastosthenes and Avienus identified her with **Isis,** the thousand-named goddess. . . clasping in her arms the young Horus, the infant southern sun-god, the last of the divine kings. This very ancient figuring appeared in the Middle Ages as the **Virgin Mary** with the child Jesus. . . Albertus Magnus of our 13th century, asserted that the Saviour's horoscope lay here. It has been said that her initials, MV, are the symbol for the sign ♍."

[4] This is one of the names of Spica. *S.N.* p.466-467: "**Spica** signifies and marks the Ear of Wheat shown in the Virgin's left hand." ". . .Hyde gave the Hebrew **Shībbōleth,** the Syrian **Shebbeltā,** the Persian **Chūshe,** and the Turkish **Salkim,** all signifying the Ear of Wheat."

Animal, the head of the Raven, and the head of the Lion.[1] Hindu astrologers say that there ascends a maiden wrapped in a cloak, and wearing worn out clothes, with a jug in her hand, and she stands in myrtle, and she wants to go to her father's house.[2] According to Ptolemy there ascends some of the Dragon's tail, the rear of the Bear and its legs and tail, and the Cup which is near the head of the Warrior, [[and part of his body]].

In the second face there ascends a figure clapping her hands and playing music. There also ascends a Man who has Half a Figure resembling the head of an ox,[3] and in his hand half of a nude man. There also ascends half of a beam with a groove in its head, and he is plowing the earth[4] with it. There also ascends the tail of the black Animal, and the middle of the Lion. Hindu astrologers say that there ascends a black man, all covered with hair. On him there are three garments, one of leather, the second of silk, and the third is a red mantle, and in his hand an inkwell in order to reckon.[5] According to Ptolemy there ascends the end of the tail of the Dragon, the tail of the

---

[1] *S.N.* pp. 254, 467 says that the ancient Arabs' Lion was a much larger constellation that extended from Gemini through Libra and parts of the constellations north and south of the zodiac. This may explain this mention of the Lion in the description of Virgo.

[2] *Picatrix:* "And there ascends in the first face of Virgo a beautiful girl covered over with a woolen sheet and holding in her hand a pomegranate. And this is a face of sowing, of plowing, of the making of trees to sprout, of gathering bunches of grapes and of the good life. And this is its form."

Agrippa: "In the first face of Virgo ascendeth the figure of a good maide, and a man casting seeds; it signifieth getting of wealth, ordering of diet, plowing, sowing, and peopling; . . ." [RH]

[3] This seems to be reference to Centaurus, which, according to *S.N.* p. 151, was also called the Half Man.

[4] Boötes, north of the feet of the Virgin, is also called the Ploughman. See *S.N.* p.92-93.

[5] *Picatrix:* "And there ascends in the second face of Virgo a man of beautiful color, dressed in leather and upon the vestment of leather another vestment of iron. And this is a face of petition, of desires, and of wealth, of tribute and the denying of things that are just. And this is its form."

Agrippa: ". . .in the second face ascendeth a black man cloathed with a skin, and a man having a bush of hair, holding a bag; they signifie gain, scraping together of wealth and covetousness." [RH]

Great Bear, and the head of the Virgin and her shoulder, and the head of the Raven with its mouth and its wings.

In the third face there ascends the remaining half of the Half Form, and the second half of the nude man, and the second half of the beam, and the tail of the Lion, and two Oxen, and half of the Herdsman. Hindu astrologers say that there ascends a white woman who is self-audatory; she is wearing a dyed mantle, her hands are leprous and she is praying to God.[1] According to Ptolemy there ascends the end of the tail of the Dragon, and the end of the Great Bear, and <the shoulder of>[2] the southern Monkey,[3] and the end of its belly, and the belly of the Raven, and the foot of the Animal.

The one born in it, his posture will be fine, his body will be erect and handsome, he will be learned and intelligent, and his hair is not curly. He loves justice, his voice is strong, he is barren, his soul is kind, his face is good looking, and he is a scribe, and he knows mathematics.

One born in the second face will be good looking, his eyes will be small, and his nostrils (nose) pretty. He is a moral person, modest, generous, and likes to be praised.

One born in the third face is also handsome, and moral, and intelligent, and modest, and wise.

In its share of the human body are the belly, the intestines, and the diaphragm, and in its share of the ailments is everything that happens to these organs, and every illness that stems from black bile. According to Egyptian astrologers the ailment of Saturn [when found in this sign] is in the genitalia, Jupiter's is in the knees, Mars' is in the belly, the Sun's is in the neck, Venus's is in the heart, Mercury's is in the chest and the Moon's is in the arms.

In its share among people are the middle ones[4], and writers, and the

---

[1] *Picatrix:* "And there ascends in the third face of Virgo a pale man of large body wrapped in a white linen cloth, and with him a woman holding in her hand black olive oil. And this is a face of weakness, old age, illness, sloth, the injury of limbs, and the destruction of the people. And this is its form."

Agrippa: "In the third face ascendeth a white woman and deaf, or an old man leaning on a staff; the signification of this is to show weakness, infirmity, loss of members, destruction of trees and depopulation of lands." [RH]

[2] Based on variant text.

[3] I was unable to identify a stellar figure by this name.

[4] Presumably neither common nor aristocratic.

learned ones, mathematicians, geometers, women, eunuchs, and those whose art brings laughter (comedians).

It is the domicile of Mercury and also its place of exaltation at 15°, and the fall of Venus at 27°, the detriment of Jupiter, and its elevation position at 23° at the present time.[1]

The rulers of the triplicity are by day Venus then the Moon, by night the Moon then Venus, and their co-ruler by day and by night is Mars.

The first face according to <the Egyptians and>[2] the Babylonians [belongs] to the Sun, the second to Venus, and the third to Mercury. According to the Hindu people the first [face belongs] to Mercury, the second to Saturn, and the third to Venus.

These are the bounds according to the Egyptians and [other] astrologers: to Mercury [are allotted] 7 degrees, to Venus 10, to Jupiter 4, to Mars 7, and to Saturn 2. According to Ptolemy to Mercury 7, and to Venus 6.[3]

The 1st and the 2nd one-ninths [belong] to Saturn, the 3rd to Jupiter, the 4th to Mars, the 5th to Venus, the 6th to Mercury, the 7th to the Moon, the 8th to the Sun, and the 9th to Mercury.

The influence of the 1st one-twelfth [belongs] to Mercury, the 2nd to Venus, the 3rd to Mars, the 4th to Jupiter, the 5th and the 6th to Saturn, the 7th to Jupiter, the 8th to Mars, the 9th to Venus, the 10th to Mercury, the 11th to the Moon, and the 12th to the Sun.

From the beginning of the sign until its end there are 5 mixed degrees, then 4 bright ones, then 2 empty ones, then 9 bright ones, then 10 mixed ones.

From the beginning of the sign until its end there are 7 female degrees, then 5 male, then 8 female, then 10 male.

The pits of the stars are the 8th, 13th, 16th, 21st, and 25th degrees. The degrees that bestow grace and honor are the 2nd, 5th, 17th, and 20th.

Of the superior stars there is the Tail of the Lion at 10° at the present time, with latitude of 11° north, of the 1st magnitude, [similar

---

[1] This is the value of the auge. See page 22, note 1.

[2] From a variant text.

[3] The Latin has "And according to Ptolemy, of Mercury 7°, of Venus 6°, of Jupiter 5°, of Saturn 6°, of Mars 6°." The Hebrew has apparently omitted the last three sets of bounds. [RH]

in nature to] the co-mixture of Saturn and Venus. There is also one star in the same degree but its latitude is 25° north, of the 2nd magnitude, [and its nature] is the co-mixture of Saturn and Venus. There are two dark stars between the Bear and the Lion, one at 11° at the present time, with latitude of 20° 40' north, and the other is in the same degree, but its latitude is 25½° north.

Libra is of the air signs, masculine, diurnal, occidental, of the winter[1] signs and changeable. At its beginning day equals night, and the night begins to grow longer and the day to grow shorter. Its [temporal hours] are shorter than the even ones. It is straight in its steps and its steps are long. Its [elemental] nature is warm and moist, but it is not a beneficial mixture, and it indicates all air and cloud mixed with winds and vapor that rises and destroys life.

On the whole it is changeable. Its beginning and middle are better than its end. When it is northern, it produces winds, and when southern, it produces moisture.

It is of human form only.[2] In its share is the center of the west and the southerly wind. Its nature is warm and moist. Its is the blood. Of the flavors, Its is the sweet, of the colors green and the earth color. Of living things, Its is man and every bird that has a big head, and of plants, its is the tall trees.

Of the regions [its are] the fifth,[3] and the land of Edom[4] from Rome to Africa, and some of the land of Kush, the sea of Barka, and Sistan, and Kabul, and Taverstan, and Balkh, and Hemdan. In its share are all cultivated [places] on top of mountains, and all soils that are not firm, and all high places,[5] and the places of market and merchandise.

---

[1] The Latin has "autumnal signs." [RH]

[2] While Libra does not have a human form, it is a product of human manufacture. Probably for this reason it has always been counted as a human sign. [RH]

[3] The fifth clima. See page 15, note 6. [RH]

[4] Edom (the Hebrew name for Idumea) which was an ancient kingdom in Biblical times, is used by Ibn Ezra also in other writings to name the Roman empire. It is not uncommon to find Edom as a synonym for an 'evil empire' in Jewish writings.

[5] The air element signifies high places, corresponding to its position in the cosmic order of elevation: fire, air, water, and earth.

Its letters are *khet* (ח) and *shin* (ש). Its years are 8 and so are its months, and the days are 20, and the hours are 16.

In the first face there ascends the figure of an angry man, in his left hand there are scales and in his right hand written books.[1] Following him there ascends a man riding on a horse and singing.[2] There ascends the head of the Dragon,[3] and the beginning of the Golden Sea,[4] and the Persian astrologers call it the Great Bear. There ascends some of the Ship.[5] According to the Hindu astrologers there ascends a man in a shop in the market with scales in his hand, and he wishes to buy and sell.[6] There ascends according to Ptolemy the middle of the tail of the Great Bear, and the middle of the body of the Virgin, her left hand where the ear of wheat is, and the tail of the Raven, and the end of the tail of the Warrior, and the end of the tail of the horse and its rear.

In the second face there ascends a Man driving a Wagon. In it there is a man with a whip in his hand and with him [another] man wearing silk clothing seated on a bed. There also ascends a small boy, and the

---

[1] As observed before, Libra, the Scales, does not contain a living form, but *S.N.* p. 272, mentions a human figure: "The sacred books of India mention it as **Tulā**, . . ., a Balance; and on the zodiac of that country it is a man bending on one knee and holding a pair of scales".

Books are not mentioned anywhere. Yet it is interesting to note that this imagery is found on the Jewish holy day of Yom Kippur, which is celebrated at the Autumn Equinox (♎) when it is believed that God writes all the deserving souls in the Book of Life. The imagery of a figure weighing on a scale a person's merits at the time of death and writing them in a book is also found in Egyptian mythology where death is associated with the west (the setting of the sun), and Libra is a western sign.

[2] This could be the image of Centaurus, lying south of the Scales.

[3] This might be the Head of the Serpent.

[4] Could not find such an image in that area.

[5] Could not find 'a ship' in this area of the sky.

[6] *Picatrix:* "And there ascends in the first face of Libra a man holding a lance in his right hand but in the left a bird hanging by its feet. And this is a face of justice, truth, good judgments, the completeness of justice of the people and weak persons, of doing good for the indigent. And this is its form."

Agrippa: "In the first face of Libra ascendeth the form of an angry man, in whose hand is a Pipe, and the form of a man reading a book; the operation of this is in justifying and helping the miserable and weak against the powerful and wicked: . . ." [RH]

middle of the Ship, and the middle of the Great Bear. There also ascends a Water Fountain.[1] Hindu astrologers say that there ascends a man in the form of an eagle, and he is naked and thirsty, and he is about to fly in the air.[2] According to Ptolemy there ascends the tail of the Dragon, and the end of the tail of the Great Bear, and the fringes of the [garment] of the Virgin.

In the third face there ascends the end of the Ship, the end of the Golden Sea, the head of the naked man with his hand on his head. There is a crown on the head of two men and each one of them has two horns on his head.[3] There also ascends according to Hindu astrologers a man whose face resembles the face of a horse, with a bow and arrows in hand.[4] There also ascends according to Ptolemy the end of the tail of the Dragon, and its arms, and its right knee, and the end of the fringes of the Virgin['s garment], and her feet.

In general a human born under this sign: his limbs will be straight and his speech pleasant, and he is a man of intelligence and morality. His hands are skilled in all arts, he can play music, he is a poet, and he likes women, and he likes to shape [things].[5] He is good hearted and

---

[1] Cannot find an image of a fountain.

[2] *Picatrix:* "And there ascends in the second face of Libra a black man having a journey of marriage and of joy. And this is a face of quiet, joy, abundance, of the good life. And this is its form."

Agrippa: ". . . in the second ascend two men furious and wrathful and a man in comely garment, sitting in a chair; and the signification of this is to shew indignation against the evil, and quietness and security of life with plenty of good things." [RH]

[3] The horns could be the Claws of the Scorpion, as Libra was "carved out" from the ancient Scorpion in the Roman era. See *S.N.* p. 269.

[4] *Picatrix:* "And there ascends in the third face of Libra a man riding upon a donkey, and before him a wolf. And this is a face of evil works, sodomites, adultery, of songs, joy, and of taste. And this is its form."

Agrippa: "In the third face ascendeth a violent man holding a bow, and before him a naked man, and also another man holding bread in one hand , and a cup of wine in the other; the signification of these is to shew wicked lusts, singings, sports and gluttony." [RH]

[5] This can be understood as 'sculpture'. The Hebrew editors have this verb in a footnote *latsur* (לצור), which means 'to shape, to give form', while in the body of the text they used the verb *latsud* (לצוד), which means to hunt. The former fits Libra better. [Additional by RH] The Latin has 'to hunt'.

generous, and his body is finer than his face, and some of them are dark looking. One born in the first face will be good looking, a wound in his head, and on his hand or leg a burn, and he is a hard worker, modest and moral.

One born in the second face will be handsome in his body, in face, and in his stature, with a defect in his eyes and his arteries, and he is charitable and friendly.

One born in the third face is also handsome in his body, with dignity in his face, and a defect in his eyes, and he is well known and honored among his people. One born at the end of the sign will be retarded or androgynous.

In its share of the human body is [the area] below the belly near the groin. Its ailments are such as stoppage of the urine, hemorrhage from below, and also darkness in the eyes.[1]

Of people in its share are the marketplace people, judges, mathematicians, musicians, and merchants who deal with food and drink.

It is the domicile of Venus, and the exaltation of Saturn at 21°, and the fall of the Sun at 19°, and the detriment house of Mars. It is the place of elevation of Mercury at 25° at the present time.[2]

According to Egyptian astrologers the ailment of Saturn [in this sign] is in the knees, Jupiter's is in the calves, Mars' is in the lower part of the belly, the Sun's is in the hands, Venus's is in the hips, Mercury's is in the heart, and the Moon's is in the chest.

The ruler of the triplicity by day is Saturn followed by Mercury, and by night it Mercury followed by Saturn, and their co-ruler by day and by night is Jupiter.

The first face according to the Egyptians and the Babylonians [belongs] to the Moon, the second to Saturn, and the third to Jupiter. According to Hindu astrologers the first [face belongs] to Venus, the second to Saturn, and the third to Mercury.

These are the bounds according to the Egyptians and [other] astrologers: [to] Saturn [are allotted] 6 degrees, Mercury 8, Jupiter 7,

---

[1] The association of eye problems, mentioned several times here, is not familiar to me as a Libra signification, nor is the arteries connection. The urine and the bleeding connection is probably from Libra governing the lower region of the belly, hence kidneys & bladder.

[2] This is the value of the auge. See page 22, note 1.

Venus 7, [and] Mars 2. According to Ptolemy Saturn [gets] 6, Venus 5, Jupiter 8, Mercury 5, and Mars 6.

The 1st one-ninth[1] [belongs] to Venus, the 2nd to Mars, the 3rd to Jupiter, the 4th and 5th to Saturn, the 6th to Jupiter, the 7th to Mars, the 8th to Venus, the 9th to Mercury, the 10th to the Moon, the 11th to the Sun, and the 12th to Mercury.

From the beginning of the sign until its end there are 5 bright degrees, then 5 mixed ones, then 8 bright ones, then 3 mixed ones, then 7 bright ones, then 2 empty ones.

From the beginning of the sign until its end there are 5 male degrees, then 5 female, then 11 male,[2] then 7 female, then 2 male.

The pits of the stars are the 1st degree, the 7th, 20th, and 30th.[3] The degrees that bestow grace and honor are 3, 5, 21.

Of the superior stars in it there is the Unarmed Leaning One called *Samach al-Azal* in the 12th degree at the present time, with southern latitude of 2°, of the 1st magnitude, [similar in nature to] the co-mixture of Venus and Mercury. There is also the Lance called Samach al-Ramach[4] in the 13th degree at the present time, with latitude of 31½° north, of the 1st magnitude, [similar in nature to] the co-mixture of Jupiter and Mars. There is a bright star at the end of the sign, with latitude of 24½° north, of the 2nd magnitude, [similar in nature to] the co-mixture of Venus and Mercury.

---

[1] The following listing of the planetary rulership within the sign goes on to 12 divisions even though the beginning of the sentence refers to the 9 parts. This could be a scribal "correction" since in the case of Libra the first nine parts have the same rulership in both divisions and the writer wished to avoid repetition.

[2] The Latin has 2 masculine, but this looks a simple error in which the Arabic numeral 11 was mistaken for the Roman numeral II. [RH]

[3] The Latin has degrees 7, 20, and 23. [RH]

[4] This is probably Arcturus in the constellation of Bootes, north of the Scales. Bootes is a herdsman usually holding a staff . *S.N.* p. 97, 100, 101: "The Staff, ultimately figured as a Lance, gave rise to the name **Al Rāmih**, which came into general use among the Arabians, but subsequently degenerated in early European astronomical works into **Aramech,** . . ." "The Arabs knew Arcturus as **Al Simāk al Rāmih**, sometimes translated the Leg of the Lance Bearer. . . **Somech Haramach** of Chilmead's *Treatise* . . ." 'Lance' in Hebrew is *romakh* (רמח) and *somekh* (סומך) means 'supporting' or even 'holding'.

Scorpio is of the water signs, female, nocturnal, northern, of the winter[1] signs and fixed. Its [temporal] hours are shorter than the even ones, it is straight in its steps and its steps are long. It indicates all [kinds of] moisture that is not steady and varies in nature, which is not useful for life but little. On the whole, [it causes] thunder and lightning. Its front part is humid and varying, its middle is mixed, and its end is stormy. When it is northern, it is humid, and when southern, it is cold. It is of the form of a scorpion. In its share is the left of the north, and the northerly wind, and the humidity that overcomes a person. Of its flavors is everything that is salty and dull; of colors its are red, green, and the color of earth. Of living things in its share are the scorpions, and [other such] creatures, and small crawling animals, and water animals, and all water that is running swiftly,[2] and all vegetation that is in water, like coral. Of plants it is the trees [and others] that are medium in height.

In its share are the third region,[3] and the land of Sheba, Arabia, Tania and Kavros. It has a share in [the rulership of] vineyards and gardens only where there is decay and places of ruin.

Its letters are *tet* (ט) and *shin* (ש), its years are 15 and so are its months, and the days are $37\frac{1}{2}$,[4] and the hours are 4.

In the first face there ascends the rear of a horse, and the rear of an ox, and a black [man] with a stick in his hand. Hindu astrologers say that there ascends the figure of a beautiful woman, her body is red and she is eating.[5] According to Ptolemy there ascends the hand of the Small Bear, and the head of the Dog, and its right arm, and the middle of the Scales.

In the second face there ascends a naked man, and the middle of a horse, and the middle of an ox. Hindu astrologers say that there ascends

---

[1] The Latin refers to Scorpio as being of the autumn. [RH]

[2] All of our texts agree on this point but this is a bit strange. With most traditional sources Scorpio rules stationary water, not flowing water. [RH]

[3] The third clima. See page 15, note 6. [RH]

[4] A footnote in the Hebrew text has 38 here, from another manuscript.

[5] *Picatrix:* "And there ascends in the first face of Scorpio a man holding a lance in his right hand, but a human head in his left. And this is a face of disposition, sadness, evil will, and hostility. And this is its form."

Agrippa: "In the first face of Scorpio ascendeth a woman of good face and habit, and two men striking her; the operations of these are for comeliness, beauty, and for strifes, treacheries, deceits, detractions, and perditions; . . ." [RH]

a woman who has left her house; she is naked and has nothing on and she is entering the sea.[1] According to Ptolemy there ascends the end of the hand of the Small Bear, the end of the Dragon's tail, the Northern Crown, the genitalia of [],[2] and his legs, and the Crown of Scorpio.

In the third face there ascends the beginning of the horse, and the carrier of the hare, and the beginning of the Ox. Hindu astrologers say that there ascends a dog, and two pigs, and a big leopard[3] with white hair, and various prey animals.[4] There ascends according to Ptolemy the figure of the Small Bear, and the foot of the One Walking on His Knees,[5] and his shoulder, and his right arm, and the belly of the Scorpion, and the head of the Censer.

One born in it will be somewhat dark, with much hair, [and] some of them will be reddish according to the aspects of the stars in the nativity. His eyes will be normal but small, and his legs are long and his feet are big, and he is fast and nimble in his walk, and his face is

---

[1] *Picatrix:* "And there ascends in the second face of Scorpio a man riding upon a camel holding a scorpion in his hand. And this is a face of knowledge, modesty, disposition, of one who speaks evilly one to another. And this is its form."

Agrippa: ". . . in the second face ascendeth a man naked, and a woman naked, and a man sitting on the earth, and before him two dogs biting one another; and their operation if for impudence, deceit, and false dealing, and for to send mischief and strife amongst men; . . ." The deviation between Ibn Ezra and these two sources begins to get quite significant from here on. [RH]

[2] The word omitted here is 'Libra'. I chose to leave it out since the phrase refers to a human body part, and Libra, the Scales, is inanimate. This could be a scribal error and the reference might be to one of the other human figures in that region of the sky — Hercules or even Ophiuchus, or to the previously mentioned human figure in the Scales. [Additional by RH] The Latin has the same reference to genitalia as the Hebrew. This suggests that if we are dealing with a scribal error, it is an old one.

[3] This could be Lupus, the Wolf. *S.N.* p. 278: ". . . **Al fahd,** the Arabian title for this constellation, their Leopard or panther;"

[4] *Picatrix:* "And there ascends in the third face of Scorpio a horse and a rabbit with it. And this is a face of evil works and taste, and joining oneself with women by force and with them being unwilling. And this is its form."

Agrippa: ". . . in the third face ascendeth a man bowed downwards upon his knees and a woman striking him with a staff, and it is the signification of drunkenness, fornication, wrath, violence, and strife." [RH]

[5] The constellation of Hercules.

large, and his forehead is narrow, and his shoulder is wide, and he is ugly, and has no voice nor pleasant speech, and he has many children. He is destructive, treacherous, irascible, liar, a gossip, depressed, generous[1] and intelligent, cunning, and deceitful.

One born in the first face will be somewhat handsome, with a mark on his head; his eyes are like the eyes of cats; his chest is wide, and some have a mark on the left leg or on the right arm, and he is a man of intelligence and talks fast.

One born in the second face, his head will be large, somewhat handsome, and he has a mark on his penis or on his back, and he is intelligent and talks much.

One born in the third face will be short in stature, his eyes are crooked, he loves to eat, he loves women, and he is depressed a lot. One born in the end of the sign is either a bastard or an imbecile.[2]

In its share of the human body are the genitalia, the hidden places, and the groin of males and females. It is [one] of the signs of deformities, indicating defect in the eye and scurvy, and the sickness that is called cancer, scabies, leprosy, pock marks and baldness, and [in] the nativity of a woman [it] is not good. The [rising] time of the 21st degree to the 24th indicates a defect in the eyes.[3]

In its share among people is every malicious and despicable man. It is the domicile of Mars, and the fall of the Moon in the 3rd degree. [The ruler of the triplicity] by day is Venus followed by Mars, and by night it is Mars followed by Venus, and their co-ruler by day and by night is the Moon.

According to Egyptian astrologers the ailment of Saturn [in it] is in the legs, Jupiter's is in the feet, Mars' is in the genitalia, the Sun's

---

[1] This attribute seems out of character here and is probably another scribal error. [Additional by RH] The Latin has *voluntariosus* here which is an undocumented form of the root *voluntar-*. Words of this root pertain to the will and are the source of the word 'voluntary'. The word could also be interpreted as 'willful' which does fit in with the rest of the adjectives in this section.

[2] Curiously enough the Latin here says "either of both or neither sex." [RH]

[3] Eye defects are associated with Antares, α Scorpii, currently at 9° of tropical Sagittarius. Antares is a name derived from Mars = Ares. See *S.N.* p. 365.

is in the heart, Venus's is in the hips,[1] Mercury's is in the stomach, and the Moon's is in the upper intestines (stomach).

According to Egyptian and Babylonian astrologers the first face [belongs] to Mars, the second to the Sun, and the third to Venus. According to Hindu astrologers the first [face belongs] to Mars, the second to Jupiter, and the third to the Moon.

These are the bounds of the Egyptians and [other] astrologers: to Mars [are allocated] 7 degrees, to Venus 4, to Mercury 8, to Jupiter 5, and to Saturn 7.[2] According to Ptolemy to Mars [are allocated] 6 degrees, to Mercury 6, to Jupiter 7, to Venus 6, and to Saturn 5.[3]

The 1st one-ninth [belongs] to the Moon, the 2nd to the Sun, the 3rd to Mercury, the 4th to Venus, the 5th to Mars, the 6th to Jupiter, the 7th and the 8th to Saturn, and the 9th to Jupiter.

The influence of the 1st one-twelfth [belongs] to Mars, the 2nd to Jupiter, the 3rd and the 4th to Saturn, the 5th to Jupiter, the 6th to Mars, the 7th to Venus, the 8th to Mercury, the 9th to the Moon, the 10th to the Sun, the 11th to Mercury, and the 12th to Venus.

From the beginning of the sign until its end there are 3 mixed degrees, then 5 bright ones, then 6 empty ones, then 6 bright ones, then 2 dark ones, then 5 bright ones, then 3 mixed ones.

From the beginning of the sign to the 4th they are male degrees, then 6 female, then 4 male, then 5 female, then 8 male, then 3 female.[4]

The pits of the stars are the 9th degree, the 10th, 17th, 22nd, 23rd and 27th degrees. The degrees that bestow grace and honor are 7, 12, and 20.

Of the superior stars there is the Scorpion's Horn in the 7th degree

---

[1] The Hebrew footnote shows the word *me'ayim* (מעיים), which means 'intestines', in some manuscripts, and the word *motnayim* (מתניים), which means 'hips' or 'waist', in others. According to the rulership scheme it should be the hips, and the other word seems a copyist error which replaced 'ת' with 'ע'. [Additional by RH] The Latin here has the 'loins'.

[2] This is an error, and should be 6. [Additional by RH] The Latin text has 6.

[3] The Latin here has the following: "6 of Mars, 7 of Venus, 8 of Jupiter, 6 of Mercury, and 3 of Saturn. [RH]

[4] The Latin has here: "4 are masculine, 6 feminine, 8 masculine, 5 feminine, 8 masculine, 3 feminine," for a total of 34 degrees. Obviously the second group of 4 masculine degrees was changed to 8. [RH]

at the present time. Its latitude is $8^1/_2$ ° north, and it is of the 2nd[1] magnitude, [similar in nature to] the co-mixture of Jupiter and Venus. There is another star called the Leg of the Animal in the 16th degree at the present time. Its latitude is 41° 10' south, and it is of the 1st magnitude, [similar in nature to] the co-mixture of Mars and Jupiter. There is the Heart of the Scorpion (Antares) in the 28th degree at the present time. Its latitude is 3° south, and it is of the 2nd magnitude,[2] [similar in nature to] the co-mixture of Mars and Jupiter.

Sagittarius is of the fire signs, male, of the diurnal signs, oriental, of the winter[3] signs, and it has two bodies. Its [temporal] hours are shorter than the even ones, and at its end the time turns around to increase the hours of the day and decrease the hours of the night in all regions. It is straight in its steps and its steps are long. Its [elemental] nature is hot and dry which is destructive to animals and plants. On the whole it is windy; its front part is wet and cold and snowing, its middle is moderate, and its end is hot. When it is northern, it is dry, and in the south it is moist. It has two forms, one half of human form and the second half of a horse form. In its share is the right of the east and the easterly wind. Its flavors are the bitter and the hot, and it produces the red bile, and its colors are everything that is yellow and reddish and the color of earth.

In its share of living things is man, and horse, and bird, and animals, and crawling creatures. In its share of the metals is lead, and of the stones it is the kind called emerald.[4]

[Its share] of the regions is the second,[5] and the land of Rai and Isphahan, and all the mountains, and it has a share in the gardens and all places of irrigation, and the place of horses and oxen, and all smooth

---

[1] A Hebrew footnote shows '1st magnitude' in another manuscript.

[2] A Hebrew footnote shows '1st magnitude' in another manuscript.

[3] The Latin has "autumnal" here. [RH]

[4] The above attributions, except for 'man', 'horse' and 'emerald' seem to be a corruption of the text, since they are usually attributed to the nearby signs, Scorpio and Capricorn. [Additional by RH] The Latin here has *plumbum* (lead) and *stagnum.* The latter means just what it looks like, non-flowing water, but may be a corruption of *stannum* which means 'tin' and tin is traditionally associated with Jupiter, the ruler of Sagittarius.

[5] The second clima. See page 15, note 6. [RH]

stones.

Its letters are *yod* (י) and *tav* (ת), its years are 12 and so are its months, and the days are 30,[1] and the hours are 12.

In the first face there ascends the figure of a Naked Man, and he is turned around with a Raven on his head.[2] There also ascends the body of a Dog,[3] and the top of a Tree.[4] Hindu astrologers say that there ascends a naked man, from his head down to his navel it is the figure of a man, and from the navel down, it is in the shape of a horse, in his hand a bow and arrows, and he is shouting.[5] There ascends, according to Ptolemy the neck of the Small Bear, the end of the tail of the Dragon, and the posterior of the One Walking on His Knees, and his nape and his head, and the end of the Scorpion, and the Knotted ones in the tail [of the Scorpion],[6] and the body of the Censer.

In the second face there ascends a figure holding in its right hand the horns of the Goat.[7] There also ascends the head of the Hyena, and

---

[1] A footnote has '33' from another manuscript.

[2] The 'Naked Man' could be Hercules, the constellation lying north of Sagittarius. *S.N.* p. 241: "Our stellar figure generally has been drawn with club and lion skin . . . but the Farnese globe shows a young man, nude and kneeling; . . ." The Raven could be Aquila, the Eagle, found in the same region of the sky. *S.N.* p. 57: "Al Achsasi, however, mentioned it as **Al Ghurāb**, the Crow, or Raven, probably a late Arabian name, and the only instance that I have seen of its application to the stars of our Aquila."

[3] The image of a dog may be found in the star α Hercules. *S.N.* p. 243: "**Ras Algethi** . . . , the Kneeler's Head;" "The nomads title for it was **Al Kalb al Rā'i**, the Shepherd's Dog, that our α shared with the adjoining *lucida* of Ophiuchus, 5° distant."

[4] Hercules holds a club in one hand and an apple branch or apple tree in the other. See *S.N.* p. 241, 242.

[5] *Picatrix:* "And there ascends in the first face of Sagittarius three bodies of men of which one is yellow, another white, but the third red. And this is a face of heat, weight, of fructifying in fields and on lands, of sustaining and dividing. And this is its form."

Agrippa: "In the first face of Sagittarius ascendeth the form of a man armed with a coat of male, and holding a naked sword in his hand; the operation of this is for boldness, malice, and liberty." [RH]

[6] Possibly the name for stars in the tail of the Scorpion. S.N. p. 372: "Al Biruni wrote that λ and υ were in the **H•arazāh,** the Joints of the Vertebrae."

[7] The only Goat that I know of in this region is Capricorn. I could not find one in connection with the Sagittarius region in the sky.

half of the Hare, and half of the Ship,[1] and the first half of the fish called Dolphin, and half of the Lizard. Hindu astrologers say that there ascends a figure of a beautiful woman with a lot of hair, wearing clothes and earrings in her ear, and in front of her there is an open chest containing golden ornaments.[2] There ascends according to Ptolemy the belly of the Small Bear, and the end of the body of the Dragon and the end of its head, and the knee of the One Walking on His Knees, and his foot and his left arm, and the end of the body <of the Animal>[3] and the Arrow and Quiver,[4] and the end of the Southern Crown.

In the third face there ascends a form of a Dog, and the end of the body of the Hyena, and the body of the Hare, and the remainder of the body of the Lion,[5] and the second half of the Ship, and the rest of the fish called Dolphin, and the tail of the Lizard, and the half of the Great Bear. Hindu astrologers say that there ascends [a figure of] a man whose color of complexion is golden, and in is hand [something that looks] like a wooden earring, and he is covered with a door made of tree bark.[6] There ascends according to Ptolemy the middle of the body

---

[1] The placement of 'Hyena', 'Hare' and 'ship' are most likely a corruption of the text, transposing them from the other side of the heavens (Lupus, Lepus & Argo), since there is no evidence for such figures in this region of the sky.

[2] *Picatrix:* "And there ascends in the second face of Sagittarius a man who leads cows, and holding [or having] a monkey and a bear before him. And this is a face of fear, lamentation, mourning, misery, and inquietude. And this is its form."

Agrippa: "In the second face ascendeth a woman weeping , and covered with cloathes; the operation of this is for sadness and fear of his [sic] own body." [RH]

[3] Based on variant text. The Serpent, the snake held by Ophiuchus. See *S.N.* p. 374, and previous footnotes in chapter 1.

[4] This might belong either to Sagittarius or the constellation Sagitta. The Arrow, nearby.

[5] No Lion is found in this region of the sky, except for the fact that Hercules is described wearing a lion's skin and head (*S.N.* p. 241). This may also be the cause for the error regarding 'Hyena', 'Hare' and 'Ship', all found near Leo, the Lion, on the other side of the heavens.

[6] *Picatrix:* "And there ascends in the third face of Sagittarius holding a cap on his heading and killing another man. And this is a face of evil inclinations, of adverse and evil effects and of swiftness in these same things and in evil inclinations, of hostility, dispersion, and of doing evilly. And this is its form."

of the Small Bear, and the end of the body of the Dragon and its head, and the end of the body of the Falling Eagle,[1] and the head of the Archer, and his shoulder, and his leg, and the Southern Crown.

A human born in it his stature will be erect, he is yellow, his testicles are long, and his legs are thick. He is a jovial, strong and generous man. His forehead is graceful and so is his beard. His hair is thin and his belly is large, and he is agile in jumping, and he loves horses, and is knowledgeable in geometry,[2] and is of [good] qualities, and he is not fixed in one way, and his voice is weak, and his children are not many.

One born in the first face will be handsome in face and appearance, and his stature will be erect, and he pursues what is goodness, and he associates with kings and magnates.

One born in the second face his body will be handsome but his face will be discolored, and his eyebrows are as if stuck together, and there is a mark on his chest.

One born in the third face will be tall, his face will be handsome, and his eyes will be like the eyes of a cat, and his chest is wide, and he is strong, and there is a mark on his left leg, and he is modest and gives advice, and [he is] an honest man. One born at the end of the sign will be lascivious. On the whole it indicates a righteous man.

In its share of the human body are the testicles,[3] and the marks, and [it indicates] a superfluous body part such as an extra finger. Of illnesses [it indicates] blindness and fever,[4] and falling from a high

---

Agrippa: "In the third face ascendeth a man like in colour to gold, or an idle man playing with a staff; and the signification of this is in following our own wills, and obstinacy in them, and in activeness for evil things, contentions, and horrible matters." [RH]

[1] As already mentioned on page 6, footnote 7 in chapter 1, this was another name for Lyra, the Lyre or the Harp. See *S.N.* p. 280, 281.

[2] This phrase can also be interpreted as 'wise in behavior'.

[3] This attribute is a bit confusing since this particular organ is usually associated with Scorpio while Sagittarius is associated with the thighs. However, it does not seem to be a mistake, since this connection appears in other places too.

[4] With a change of one letter, baldness (*karakhat* קרחת) can be fever (*kadakhat* קדחת), which seems more appropriate here. [Additional by RH] The Latin text supports the reading of fever which the Levy and Cantera translation also has.

place, and the illnesses that come from the venom of animals and snakes, and [those that come] from the severance of a limb. From its 15th degree to the 18th it indicates a defect in the eyes.

In its share among people are the judges, the worshippers of God, the philanthropists and people of mercy, the interpreters of dreams, the archers, and the merchants.

It is the domicile of Jupiter, the exaltation of the South Node at 3°, the detriment of Mercury, and the place of elevation of Saturn in the 12th degree at the present time, and the place of the nadir of the Sun and Venus at 27° at the present time.[1]

According to Egyptian astrologers the ailment of Saturn [in it] is in the feet, Jupiter's is in the head, Mars' is in the testicles, the Sun's is in the heart, Venus's is in the genitalia, Mercury's is in the hips, and the Moon's is in the belly.

The ruler of the triplicity by day is the Sun followed by Jupiter, and by night it is Jupiter followed by the Sun, and their co-ruler by day and by night is Saturn.

The first face according to the Egyptians and the Babylonians [belongs] to Mercury, the second to the Moon, and the third to Saturn. According to Hindu astrologers the first [belongs] to Jupiter, the second to Mars, and the third to the Sun.

These are the bounds of the Egyptians and [other] astrologers: to Jupiter [are allocated] 12 degrees, to Venus 5, to Mercury 4, to Saturn 5, and to Mars 4. According to Ptolemy to Jupiter [are allocated] 8 degrees,[2] to Venus 6, to Mercury 5, to Saturn 6, and to Mars 5.

The 1st one-ninth [belongs] to Mars, the 2nd to Venus, the 3rd to Mercury, the 4th to the Moon, the 5th to the Sun, the 6th to Mercury, the 7th to Venus, the 8th to Mars, and the 9th to Jupiter.

The influence of the 1st one-twelfth [belongs] to Jupiter, the 2nd and the 3rd to Saturn, the 4th to Jupiter, the 5th to Mars, the 6th to Venus, the 7th to Mercury, the 8th to the Moon, the 9th to the Sun, the 10th to Mercury, the 11th to Venus, and the 12th to Mars.

From the beginning of the sign until its end there are 9 bright degrees, then 3 mixed ones, then 7 bright ones, then 4 dark ones, then

---

[1] These are values of the auges or oppositions of the auges. See page 22, note 1.

[2] The Latin has 4 here which is clearly wrong. [RH]

7 empty ones.[1]

From the beginning of the sign until its end there are 2 male degrees, then 3 female, then 7 male, then 12 female, then 6 male.

The pits of the stars are the 7th degree, the 12th, 15th, 24th, 27th, and the 30th one. The degrees that bestow grace and honor are the 13th and the 20th.

Of the dark[2] stars there is the one that is after the tail of the Scorpion in the 17th degree at the present time, with latitude of 13° 15' south, and the star called the Arrow[3] in the 18th degree[4] at the present time, with latitude of 6° 20' south at the present time. There is the star called the Eye of the Archer[5] which is at the end of the sign at the present time, with latitude of 45' north.

Capricorn is of the earth signs, female, of the nocturnal signs, southern, of the cold season signs, and it is changeable, and in it the hours of the day begin to increase and the hours of the night to decrease. Its [temporal] hours are shorter than the even ones. Its [elemental] nature is cold and dry, and harmful, and its front part is warm and moist, its middle is mixed, and its end is rainy whether northern or southern. It is crooked in its steps and its steps are short. It is missing [a part] in its eyes,[6] and it has two forms and two elements. Its first half is of the form of land animals that have hoofs, and its second half is of the form

---

[1] The Latin has here "9 degrees are bright, 9 mixed or smokey, 7 bright, 4 dark, and 7 mixed," for a grand total of 36 degrees. Obviously the Latin text is corrupt here. [RH]

[2] The Hebrew text has the word *hashuhim* (חשוכים) which means 'dark', but it seems to me that this might have been *hashuvim* (חשובים) which means 'important'. [Additional by RH] For what it is worth the Latin also has 'dark' here which again means that if there is a corruption of the text, it is very old.

[3] This might be γ Sagittarius. *S.N.* p. 357: "**Al Nasl**, the Point .... Marking the head of the arrow . . ."

[4] The Hebrew footnote has 'the 28th degree' in some manuscripts.

[5] *S.N.* p. 359: "ν^1 and ν2 . . . were **Ain Al Rāmī**, the Archer's Eye."

[6] This attribute, which I could not find elsewhere, might be explained by comment in *S.N.* p. 136, regarding Capricornus: "Eratosthenes knew it as Πᾶν and 'Αιγι-Παν, the Goat-Footed Pan, *half finishes* [emphasis mine], Smith said, by his plunge into the Nile in panic at the approach of the monster Typhon."

of aquatic animals.[1]

In its share is the black bile.[2] Its flavors are the astringent and bitter, and its colors are black and the color of earth.

In its share of living [things are] all quadrupeds with hoofs, and some aquatic animals. [Also] in its share are the worms, the fleas, and the flies, and of the plants [it has] the olives, nuts, carobs, gall nuts, every tree that has many thorns, and [vegetation that grows] around lakes like bamboo and reed.

In its share of the regions it has the first one,[3] and the land of Kush, and Makran, and Amman, and Sind, and Alhind, and Alhodu (India), and the border area of Idumea.

In its share are the orchards, all irrigated places, fountain heads, rivers and reservoirs, the dens of dogs and foxes, houses of prison and of slaves, and the place of fire after it has been extinguished, and the pasture ground of the sheep, and every place where nothing grows.[4]

Its letters are *kaf* (ב) and *het* (ח). Its years are 27[5] and so are its months, and the days are 307½, and the hours are 14.

In the first face there ascends the second half of the Great Bear, and the figure of a woman playing music,[6] and the head of a big Fish,[7] and a Fountain of Bad Waters, and the body of a monkey, and the head of a dog. Hindu astrologers say that there ascends an irascible black man, his body is like that of a wild boar, with much hair, and his teeth

---

[1] Capricorn is depicted as half goat and half fish.

[2] Black bile is the Hebrew expression for 'depression' which is a known Capricorn trait. Earth signs in general are melancholic.

[3] The first clima. See page 15, note 6. [RH]

[4] There is an obvious contradiction here, which can either be a corruption of the text, or it may be that this is due to the dual nature of Capricorn — earth (dryness) from the goat part and water (for orchards and the pasture) from the fish part. [Additional by RH] Lilly refers to Capricorn as ruling "fallow places." These are not places where nothing *can* grow, but rather places where no crop *is* growing.

[5] In the case of every other sign the years are the same as the least period of the ruling planet which in this case would be 30. However, Vettius Valens also seems to give 27 years to Capricorn in his planetary period systems. See Vettius Valens, *Anthology,* Book IV, chapter X. [RH]

[6] This is probably Lyra even though this constellation does not contain a human figure.

[7] Delphinus, possibly.

are sharp and as long as beams, and he has a cattle goad, and he catches fish.[1]

There ascends according to Ptolemy the middle of the body of the Small Bear and its neck, and the end of the body of the Falling Eagle.

In the second face there ascends a Woman Sitting on a Bed, with her Vine,[2] and a big Fish, and half a Wagon. Hindu astrologers say that there ascends a black woman covered with a mantle and she has a horse.[3] There ascends according to Ptolemy the end of the Small Bear, and the end of the body of the Dragon, and the right foot of the Hen,[4] and her neck, and her head, and the body of the Flying Eagle, and the horns of Capricornus, and its head, and the edges of the Archer.

In the third face there ascends the tail of the Fish, the end of the Bad [Waters] Fountain, and the end of the Monkey, the second half of the Wagon, and half a Figure Without a Head because its head is in its hand. Hindu astrologers say that there ascends a beautiful woman, though black, and her hands are skilled in all kinds of work and spinning of silk.[5] There also ascends according to Ptolemy the end of

---

[1] *Picatrix:* "And there ascends in the first face of Capricorn holding a pipe in his right hand but a hoe in his left hand. And this is a face of happiness, joy, and the scattering of tasks and laziness with weakness and unceasing evils. And this is its form."

Agrippa: "In the first face of Capricorn ascendeth the form of a woman, and a man carrying full bags; and the signification of these is to go forth and rejoyce, to gain and to lose with weakness and baseness: . . ." [RH]

[2] The Hebrew word is *gefen* (גפן), but the footnote has an alternative word *sapan* (ספן), which can be read to mean 'sailor'.

[3] *Picatrix:* "And there ascends in the second face of Capricorn a man having before him half of a monkey. And this is a face of seeking matters which can in no way be, nor does anything prevail to touch upon these matters. And this is its form."

Agrippa: ". . . in the second face ascendeth two women, and a man looking towards a Bird flying in the Air; and the signification of these is for the requiring those things which cannot be done, and for the searching after those things which cannot be known: . . ." [RH]

[4] Cygnus.

[5] Vega, in Lyra, is one of the stars in the Chinese figure of the Spinning Damsel or the Weaving Sister. See *S.N.* p. 285. [Additional by RH]

Picatrix: "And there ascends in the third face of Capricorn a man holding a book and opening and closing it, and having before the book the tail of a fish.

the Dragon, and the end of the body of the Hen, and its right foot, and its left wing, and the fish called Dolphin, and the middle of the body of Capricornus, and the tail of the Fish.

A human born in it his body will be handsome, his posture erect, but he is dry [looking] and his head is small, and his cheeks are thick and his beard is impressive; there is no hair on his chest, and his voice is thin. He is irascible, and destructive, and clever, and cunning, and worries a lot, and loves intercourse and fornication. He has many children and twins. His endeavors are unsuccessful and he has little strength. He will have wealth through kings, and a bad experience will befall him because of women.

One born in the first face his body will be fine, his chest wide and a black mark under his arm, and he is knowledgeable and modest, and intelligent and generous.

One born in the second face will also be handsome in body, and his nostrils are long, and his eyes good looking, and his nature is evil, and he is irascible [yet] friendly.

One born in the third face is also handsome in body, but his face is discolored, and he has a mark on his left arm or on his testicle. He is easy to anger, abhors evil, loves women, intelligent and friendly. One born at the end of the sign is a bastard. In general this sign is not good in the nativity of women.

In its share of the human body are the testicles and all the diseases that befall in them. Of the diseases [in its share] are scabies, pruritus, leprosy, muteness and deafness, baldness,[1] blindness and hemorrhage from below. From 22 to 25 degrees it indicates a defect in the eye.

In its share among people are the farmers, the sailors, the middle

---

And this is a face of riches, the accumulation of money and the ascent of business affairs tending toward a good end. And this is its form."

Agrippa: ". . . In the third face ascendeth a woman chast in body, and wise in her work, and a banker gathering his mony together on the table; the signification of this is to govern in prudence, in covetousness of money, and in avarice."

[1] With a change of one letter, 'baldness' (*karakhat* קרחת) can be 'fever' (*kadakhat* קדחת), which seems more appropriate here. [Additional by RH] The Latin text supports the reading of 'fever' which the Levy and Cantera translation also has.

ones,[1] and the shepherds.

It is the domicile of Saturn, and the exaltation of Mars at 28 degrees, and the fall of Jupiter at 15 degrees, and the detriment of the Moon, and there is the north node of Mercury at 26 degrees.

The rulers of the triplicity by day are Venus followed by the Moon, and by night the Moon followed by Venus, and their co-ruler by day and by night is Mars.

According to the Egyptians the ailment of Saturn [in it] is in the head, Jupiter's is in the neck, Mars' is in the knees, the Sun's is in the belly, Venus's is in the testicles, Mercury's is in the genitalia, and the Moon's is in the hips.[2]

The first face according to the Egyptians and the Babylonians [belongs] to Jupiter, the second to Mars, and the third to the Sun. According to the Hindu astrologers the first [face belongs] to Saturn, the second to Venus, and the third to Mercury.

These are the bounds of the Egyptians and [other] astrologers: Mercury [gets] 7°, Jupiter 7°, Venus 8°, Saturn 4°, and Mars 4°. According to Ptolemy Venus [gets] 6°, Mercury 6°,[3] Jupiter 7°, Saturn 6°, and Mars 5°.

The 1st one-ninth [belongs] to Saturn and so does the 2nd, the 3rd to Jupiter, the 4th to Mars, the 5th to Venus, the 6th to Mercury, the 7th to the Moon, the 8th to the Sun, and the 9th to Mercury.

The 10th[4] to Venus, the 11th to Mars, and the 12th to Jupiter.

From the beginning of the sign until its end there are 7 mixed degrees, then 3 bright ones, then 5 dark ones, then 4 bright ones, then 2 mixed ones, then 4 empty ones, and then 5 bright ones.

From the beginning of the sign until its end there are 11 male degrees, then 8 female ones, then 11 male ones.

The pits of the stars are the 2nd degree, the 17th, 22nd, 24th, and

---

[1] Common or middle class.

[2] The text has *me'ayim* (מעיים) which means 'intestines', but 'hips' is the correct word, as was explained previously in footnote 1, page 58.

[3] In the Latin Mercury get 8 which is clearly wrong. [RH]

[4] Ibn Ezra is lumping together the division of 9 and 12, since, as in the case of Libra, the rulership sequence within the sign for the 9 and the 12 divisions overlap for the first nine parts. [Additional by RH] The Latin text lists them separately.

28th. The degrees that bestow grace and honor are 13,[1] 14, and 20.

Of the superior stars there are the Falling Eagle,[2] at 3° at the present time, with latitude of 62° north, of the first magnitude, [similar in nature to] the co-mixture of Venus and Mercury. There is [also] the Flying Eagle[3] at 19° at the present time, with latitude of 29° 10' north, of the second magnitude, of the nature of Jupiter. There are cloudy stars from 7° to 13°.

Aquarius is of the air signs, male, of the diurnal signs, occidental, of the cold [season] signs, [and] fixed. Its [temporal] hours are shorter than the even ones, it is crooked in its steps and its steps are short. Its [elemental] nature is warm and moist, and destructive, and it indicates all air that is harmful to life, and all winds that bring loss and destruction. Its front part is very damp, its middle is mixed, and its end produces winds. When it is northern, it produces snow, and when southern, it produces clouds.

It is of human form only. In its share is the left of the west, and its is the sea-wind. In its share is the blood, and its flavor is sweet, and its colors are the green, saffron, and the color of dust. In its share are people, and ministers,[4] and every ugly man.

In its share of the regions are the second,[5] and the land of Kush, Al-Kufa, Al-Haviza, and Al-Kabat.

In its share are the running water, and the seas, and the places of distilleries,[6] and places where wine is sold, and every land that is in the mountains, and all [places of] drink, and brothels, and every vessel that

---

[1] The Latin adds the 12th degree. [RH]

[2] Vega, α Lyra.

[3] α Aquila.

[4] The Hebrew word *sarim* (שרים) means ministers or princes, but, based on another description by Ibn Ezra in the *Book of Reasons* regarding Aquarius, it could be read as *shedim* (שדים) meaning devils or ghosts. [Additional by RH] The Latin reads as follows: "... the prince and every lay person retained among them."

[5] The second clima, but see page 15, note 6. [RH]

[6] The word is *zehuhit* (זכוכית) which normally means 'glass'. The root is also used in words that mean 'pure' or 'to purify'. 'Glass' seems out of the usual context of Aquarius while 'distilleries' fits the phrase that follows that relates to wine.

is used to draw water.[1]

Its letters are *lamed* (ל) and *dalet* (ד). Its years are 30 and so are its months, and the days are 75, and the hours are 6.

In the first face there ascends the head of the one in whose hand is the Horse, and a bird whose head is black, and it catches fish.

Hindu astrologers say that there ascends a black man who is skilled in copper.[2] According to Ptolemy there ascends the end of the tail of the Small Bear, and the foot of the Hen, and the head of the first Horse, and the posterior of the Goat and its tail.

In the second face there ascends the body of the Horse and the wing of the bird that catches fish. Hindu astrologers say that there ascends a very black man whose beard is long and in his hand a bow and arrows, and purses that contain precious stones and gold.[3] There ascends according to Ptolemy the tail of the Small Bear, and the body of the Second Horse,[4] and the first, and the beginning of the Bucket,[5] and the middle of the belly of the Southern Fish.

In the third face there ascends the Hen, and the rear of the One

---

[1] The Latin has "Of places it has [those] of running water, and the places of seas and of canals, and places of the selling of wine, and every mountainous [reading *montuosam* for *mortuosam*] land, and every place of irrigation, and of the arranging of pimps, and of every vessel in which water is hauled."

[2] *Picatrix:* "And there ascends in the first face of Aquarius a man who has his head cut short [off?] and who holds in his hand a peacock. And this is a face of misery, poverty, and of a slave who deals with crumbs. And this is its form."

Agrippa: "In the first face of Aquarius ascendeth the form of a prudent man, and of a women spinning; and the signification of these is in the thought and the labour for gain, in poverty and baseness . . ." [RH]

[3] *Picatrix:* "And there ascends in the second face of Aquarius a man like unto a king who values himself much and who shuns those whom he sees. And this is a face of beauty and of position, of having that which one seeks, of completeness, of harm and of weakness. And this is its form."

Agrippa: ". . . in the second face ascendeth the form of a man with a long beard; and the signification of this belongeth to the understanding, meekness, modesty, liberty and good maners: . . ." [RH]

[4] *S.N.* p. 324, regarding Pegasus: "Arabian astronomers followed Ptolemy in Al Faras al Thānī, the Second Horse. . ." The first Horse is probably Equus.

[5] According to *S.N.* p 45, Arab astrologers call the whole constellation of Aquarius 'Bucket'.

who is holding the Horse, and the end of the Bird that catches fish. Hindu astrologers say that there ascends an angry and deceitful black man who has hair in his ear, and on him a crown from the leaves of a tree, and he turns from place to place.[1] There ascends according to Ptolemy the tail of the Small Bear, and the body of the Horse, and the end of the Bucket, and the head of the Southern Fish.

A person born in it will be short, his head will be large, with one leg thicker than the other. He is generous, handsome and self-laudatory, and all he desires is to increase wealth, and he is barren or has few children.

One born in the first face will be handsome in body and face, and he has a mark on his chest or on his left leg. He is intelligent and gregarious.

One born in the second face will be tall, with ruddy face, with a mark on his back and under his arms, and all his days are [lived] in hardship.[2]

One born in the third face will be short, handsome in body and face, and he is ruddy, with a mark under his arm, and he likes women. One born at the end of the sign will be different in appearance and in all his actions.

In its share of the human body are the legs and all the ailments that occur in them, and the black bile (depression), and the black jaundice, and the obstruction of the arteries. From 20 degrees to 25 it indicates a defect in the eyes.

In its share among people is every lowly and afflicted person, and tillers of the land, and the tanners.

It is the domicile of Saturn and the detriment of the Sun, and has no dignity nor debility for any other star. It is the low position of Mars at 12° at the present time.[3]

---

[1] *Picatrix*: "And there ascends in the third face of Aquarius a man with [his] head shortened [cut off?] and who has an old woman with him. And this is a face of abundance, of the perfection of the will, and of insulting behavior. And this is its form."

Agrippa: ". . . in the third face ascendeth a black and angry man; and the signification of this is in expressing insolence, and impudence." [RH]

[2] The word can also be understood as 'grief'. [Additional by RH] The Latin has '*dolore*' which is 'with pain' or 'with grief'.

[3] This is the value of the opposition of the auge. See page 22, note 1.

According to the Egyptians the ailment of Saturn [in this sign] is in the neck, Jupiter's is in the arms, Mars' is in the legs, the Sun's is in the intestines, Venus's is in the knees, Mercury's is in the testicles, and the Moon's is in the genitalia.

The rulers of the triplicity by day are Saturn followed by Mercury, by night Mercury followed by Saturn, and their co-ruler by day and by night is Jupiter.

The first face according to the Egyptians and the Babylonians [belongs] to Venus, the second to Mercury, and the third to the Moon. According to Hindu astrologers the first [face belongs] to Saturn, the second to Mercury, and the third to Venus.

These are the bounds according to [the] Egyptians and [other] astrologers: to Mercury [are allocated] 7 degrees, to Venus 6, to Jupiter 7, to Mars 5, and to Saturn 5. According to Ptolemy to Saturn [are allocated] 6 [degrees], to Mercury 6, to Venus 8 to Jupiter 5, and to Mars 5.

The 1st one-ninth [belongs] to Venus, the 2nd to Mars, the 3rd to Jupiter, the 4th and the 5th Saturn, the 6th to Jupiter, the 7th to Mars, the 8th to Venus, and the 9th to Mercury.

The influence of the 1st one-twelfth [belongs] to Saturn, the 2nd to Jupiter, the 3rd to Mars, the 4th to Venus, the 5th to Mercury, the 6th to the Moon, the 7th to the Sun, the 8th to Mercury, the 9th to Venus, the 10th to Mars, the 11th to Jupiter, and the 12th to Saturn.

From the beginning of the sign until its end [there are] 4 dark degrees, then 5 bright ones, then 4 mixed ones, then 8 bright ones, then 4 empty ones, then 5 bright ones.

From the beginning of the sign until [its] end [there are] 5 male degrees, then 7 female, 6 male, 7 female, and 5 male ones.

The pits of the stars are the 1st degree, the 12th, 17th, 22nd,[1] and the 29th. The degrees that bestow grace and honor are 9, 17, and 20.[2]

Of the superior stars there is the Mouth of the Southern Fish[3] at 23° at the present time, with latitude of 23° 20' south, of the first magnitude, [similar in nature to] the co-mixture of Venus and Mercury.

---

[1] The Latin has the 23rd instead of the 22nd. [RH]

[2] The Latin has 7, 16, 17, and 20. [RH]

[3] Fomalhaut.

Another star called *Al-Ridf*[1] in the tail of the Hen at 25° at the present time, with latitude of 60° north, of the second magnitude, [similar in nature to] the co-mixture of Venus and Mercury. There are also four dim stars in the head of the Horse between 12 and 14 degrees.

Pisces is of the water signs, female, nocturnal, northern, of the cold [season] signs, and it has two bodies. At its end day equals night; its [temporal] hours are shorter than the even ones; it is crooked in its steps and its steps are short. Its limbs are severed,[2] and it produces cold and moisture that is destructive to the living and the plants, and [it indicates] all foul waters. On the whole it increases winds. Its front part is moderate, its middle is cold, and its end is somewhat warm. When it is northern, it produces winds, and when southern, it produces water. It is of the form of a [[fish]].

In its share is the right of the north, and the northern wind which is between north and east. Its [elemental] nature is cold and moist, and it produces the bad phlegm in the human body. Its flavors are the salty and the insipid; its colors are green, and white, and variable colors.

In its share of living [things] are the aquatic animals, and all vegetation that is in the water, and crystal, coral and onyx.

In its share of the regions is the second one,[3] and the land of Sheba, and Tabaristan, and the land north of Gurgan, and the co-rulership of Idumea and Alexandria.

In its share [of places] are synagogues, and the banks of all rivers and lakes.

Its letters are *mem* (מ) and *tsade* (צ); its years are 12 and so are its months, and the days are 30, and the hours are 12.

He is mute and has many children.

In the first face there ascends half a Horse that has Wings, and the beginning of the River,[4] and the tail of the fish called Crocodile.[5] Hindu

---

[1] This is another name for α Cygnus, Deneb, the Hen's Tail, as is described in *S.N.* p. 195. The Hebrew footnote has an alternative name, Al-Nadaf, from other manuscripts.

[2] Due to the short ascension. [Additional by RH] This attribution *is* generally given to Pisces but the Latin says, "but its limbs are whole." Apparently a 'not' got omitted somewhere on the way to the Latin text.

[3] The second clima. See page 15, note 6. [RH]

[4] Eridanus.

[5] Cetus.

astrologers say that in the first face there ascends a man wearing beautiful clothes, in his hand an iron instrument, and he is going home.[1] There ascends according to Ptolemy the end of the tail of the Small Bear, the belly of the Second Horse, and the beginning of the First Fish.

In the second face there ascends half of the River, and the second half of the One Walking on His Knees.[2] Hindu astrologers say that there ascends a beautiful white woman, sitting in a ship at sea, and she wishes to go out on land.[3] There ascends according to Ptolemy the tail of the Small bear, and the hand of the Woman Sitting on a Chair,[4] and the shoulder of the Woman who has no Husband,[5] and the Head of the Woman that is Joined with the Horse,[6] and the latter part of the First Fish.

In the third face there ascends the end of the River, and the end of the Fish called Crocodile, and the remainder of the One Walking on His Knees. Hindu astrologers say that there ascends a naked man putting his foot on his belly, in his hand a lance, and he is shouting out of fear of

---

[1] *Picatrix:* "And there ascends in the first face of Pisces a man who has two bodies and is like he is going to salute with his hands. And this is a face of peace and humility, of weakness, of many journeys, of misery, if seeking wealth, and the lamenting of [one's] manner of living. And this is its form."

Agrippa: "In the first face of Pisces ascendeth a man carrying burthens on his shoulder, and well cloathed; it hath his signification in journeys, change of place, and in carefulness of getting wealth and cloaths: . . ."

[2] This placement is puzzling, since this description belongs to Hercules, which is not even near this region of the sky. The figure is mentioned again below, in the section about the third face.

[3] *Picatrix:* "And there ascends in the second face of Pisces a man turned around backwards holding holding his head downwards, and [his] feet upwards lifted up on high, and in his hand a tray of something to be eaten. And this is a face of reward and of strong will in matters which are high, burdensome and valued, and of thinking. And this is its form."

Agrippa: ". . .in the second face ascendeth a woman of a good countenance, and well adorned; and the signification is to desire and put ones self on about high and great matters: . . ." [RH]

[4] Cassiopeia.

[5] Andromeda.

[6] The head of Andromeda is also part of Pegasus, the Horse.

robbers and fire.[1] There ascends according to Ptolemy the end of the tail of the Small Bear, and the middle of the back of the Woman Sitting on a Chair, and the chest of the Woman Who Has No Husband, and some of the Flaxen Thread, and the rear of the Serpent.[2]

A human born in it will be erect with medium [size] body. His complexion is white, his chest is wide, his beard is graceful, his forehead is clear, and his eyes have more black than white. There are some that are missing a limb. He likes sleep, and he is a glutton, and he is not stable, and he is irascible, intelligent, deceitful, and has a weak voice.

One born in the first face will be handsome in his body and his face, and his chest is wide, and he has a mark under his arm or under his legs.

One born in the second face will be short, handsome, his beard is black, he is hairy, has a mark under his arm, and he is contentious with all people.

One born in the third face will be yellow, with beautiful eyes, and [will suffer from] many illnesses. One born at the end of the sign will kill himself.

In its share of the human body are the feet and the toes and all the ailments that occur in them. It causes the plague, skin disease, leprosy, and in general it is the sign of illnesses.

In its share among people are the lowly ones and fishermen.

The Egyptians said that the ailment of Saturn in it is in the arms, Jupiter's is in the upper bowels (stomach), Mars' is in the feet, the Sun's is in the genitalia, Venus's is in the legs, Mercury's is in the knees, and the Moon's is in the testicles.[3]

---

[1] *Picatrix:* "And there ascends in the third face of Pisces a sad man and of evil thoughts thinking on deceptions and treachery; and before him a woman and an ass ascending over her, and in her hand a bird. And this is a face of ambition and of lying with women with a great appetite, and of seeking quiet and peace. And this is its form."

Agrippa: ". . . in the third face ascendeth a man naked, or a youth, and nigh him a beautiful maide, whose head is adorned with flowers, and it hath his signification for rest, idleness, delight, fornication, and for imbracings of women." [RH]

[2] Possibly Cetus, the Sea Monster.

[3] The logic of this system should make the Moon's rulership over the thighs not the testicles.

It is the domicile of Jupiter, the exaltation of Venus at 27°, the detriment of Mercury at 15° and [also] its fall. It is the low position of Jupiter at 23° at the present time,[1] and the north node of Venus at 26° at the present time. The rulers of the triplicity by day are Venus followed by Mars, by night it is the reverse, and their co-ruler by day and by night is the Moon.

The first face according to the Egyptians and the Babylonians [belongs] to Saturn, the second to Jupiter, and the third to Mars. According to Hindu astrologers the first [face belongs] to Jupiter, the second to the Moon, and the third to Mars.

These are the bounds of the Egyptians and [other] astrologers: to Venus [are allocated] 12 degrees, to Jupiter 4, to Mercury 3, to Mars 9, and to Saturn 2 degrees. According to Ptolemy to Venus 8, to Jupiter 6, to Mercury 6, to Mars 5, and to Saturn 5.

The 1st one-ninth [belongs] to the Moon, the 2nd to the Sun, the 3rd to Mercury, the 4th to Venus, the 5th to Mars, the 6th to Jupiter, the 7th and the 8th to Saturn, and the 9th to Jupiter.

The influence of the 1st one-twelfth [belongs] to Jupiter, the 2nd to Mars, the 3rd to Venus, the 4th to mercury, the 5th to the Moon, the 6th to the Sun, the 7th to Mercury, the 8th to Venus, the 9th to Mars, the 10th to Jupiter, and the 11th, and the 12th to Saturn.

From the beginning of the sign until its end there are 6 bright degrees,[2] then 6 mixed ones, then 4 bright ones, then 3 empty ones, then 3 bright ones, then 2 mixed ones.

From the beginning of the sign until its end there are 10 male degrees, then 10 female, then 3 male, then 3 female, then 2 male.[3]

The pits of the stars are in the 4th degree, the 9th, 24th, 27th, and 28th degrees. The degrees that bestow grace and honor are 18 and 20.[4]

Of the superior stars there is the Horse's Shoulder[5] at 18° at the present time, with latitude of 31° north, of the second magnitude, [similar in nature to] the co-mixture of Mars and Mercury.

---

[1] This is the value of the opposition of the auge. See page 22, note 1.

[2] The Latin here has 6 smokey degrees. [RH]

[3] These do not add up to 30. The Latin here has 10 masculine, 10 feminine, 3 masculine, 5 feminine, and 2 masculine which does add up to 30. [RH]

[4] The Latin 13 and 28. [RH]

[5] Markab, α Pegasus.

The Superior [Fixed] Stars that I have mentioned when they appear in the nativity of a person in the ascending degree,[1] or at the Midheaven degree which is the beginning of the 10th house,[2] or in the same degree of the Sun by day, or with the Moon by night, or with the degree of the Lot of the Moon which is called the Good Fortune,[3] then the person will rise to a high level such as his forebears had never reached nor that anyone could have imagined. Yet all the ancients agree with the opinion that this individual will meet with unfortunate end, especially if the [star] indicating this is one of the stars whose is nature of the co-mixture of the malefics planets.

Now I shall discuss the nature of the stars that are in the ecliptic, as well as the northern and southern constellations.

The ancients said that the stars in the mouth of the Ram are of the nature of Mercury and somewhat of Saturn. The star on its leg is of the nature of Mars and the one on the rump is of the nature of Venus.

The stars in the place where the Bull is severed are of the nature of Venus and somewhat of Jupiter. The Pleiades are [similar in nature to] the co-mixture of Mars and the Moon. The stars on the head of the Bull are of the nature of Saturn and somewhat of Mercury; those include the great star, the Eye of the Bull, called Aldebaran, which is of the nature of Mars alone. Those on the horns of the Bull are also of that nature.

The [stars on the] foot of the Twins are of the nature of Mercury and somewhat of Venus. Those in the testicles of the Twins are of the nature of Saturn. The two bright stars in the head of the first [Twin] is of the nature of Mercury, and the second is of the nature of Mars.

The stars on the feet of the Crab are of the nature of Mercury and somewhat of Mars. Those in its tail are [similar in nature to] the co-mixture of Saturn and Mercury. Those on its belly, the Cloudy Ones,

---

[1] The phrase "in the ascending degree" raises the question of whether a fixed star has to be bodily on the horizon, or in an ecliptic conjunction with the ascending degree. These two ways of regarding a star's position can lead to very different results depending upon the star's latitude. The wording here supports the latter practice. [RH]

[2] This is clear and unambiguous evidence that Ibn Ezra used a quadrant system of houses of the modern type in which the Ascendant is the cusp of the first house and the Midheaven the cusp of the tenth house. [RH]

[3] Also known as the Lot or Part of Fortune. [RH]

are [similar in nature to] the co-mixture of Mars and the Moon. <Those leaning on its back, called the Heroes,[1] are of the nature of Mars and the Sun.>[2]

Those in the head of the Lion are of the nature of Saturn and somewhat of Mars. Those in its neck are of the nature of Saturn and somewhat of Mercury. The bright star called the Lion's Heart is [similar in nature to] the co-mixture of Mars and Saturn. The one in its loins is of the nature of Venus, and the ones in its testicles are of the nature of Venus and somewhat of Mercury.

The stars over the head of the Virgin on the tip of her right wing are of the nature of Mercury and somewhat of Mars. Those on the hips are of the nature of Venus and those on the left wing are [similar in nature to] the co-mixture of Saturn and Mercury. The star called *Simak Al-Azel*[3] is of the nature of Venus and somewhat of Mercury. The stars on her feet are of the nature of Mercury and somewhat of Mars.

The two stars on the Scales are [similar in nature to] the co-mixture of Jupiter and Mercury.

The star on the horns of the Scorpion is of the nature of Saturn and somewhat of Mercury and Mars. The bright stars on the back of the Scorpion are of the nature of Mars and somewhat of Saturn. The Scorpion's Heart is of the nature of Mars and somewhat of Jupiter. The ones in its tail are of the nature of Saturn and somewhat of Venus. The cloudy stars in it are [similar in nature to] the co-mixture of Mars and the Moon, and the cloudy [stars] in the edges are [similar in nature to] the co-mixture of Mercury and the Moon.

The stars on the back of the Archer are of the nature of Jupiter and somewhat of Mercury. The legs of the horse are [similar in nature to] the co-mixture of Jupiter and Saturn. The stars in its tail are of the nature of Venus and somewhat of Saturn.

The stars in the mouth of the Goat are of the nature of Mars and somewhat of Venus. Those on its belly are [similar in nature to] the co-mixture of Mars and Mercury, and those on its tail are [similar in nature to] the co-mixture of Saturn and Mercury.

---

[1] The Hebrew word is *gibborim* (גבורים), which means 'heroes', but it looks to me more like a distortion of *dvorim* (דבורים), which means 'bees', which fits the nebula called Beehive in the constellation of the Crab.

[2] The text in <> is missing in the Latin and in Levy & Cantera.

[3] Spica. [RH]

The stars on the shoulder of the Bucket Bearer and his left arm are [similar in nature to] the co-mixture of Saturn and Mercury. Those on his nostrils are of the nature of Mercury and somewhat Saturn. Those on the flow of the water are of the nature of Saturn and somewhat of Jupiter.

[The stars] on the head of the Fishes are of the nature of Mercury and somewhat of Saturn. [Those] on its belly are [similar in nature to] the co-mixture of Jupiter and Mercury. Those on its tail are of the nature of Saturn and somewhat of Mercury. Those on the northern side of the belly are of the nature of Jupiter and somewhat of Venus, and those in the end of the Fishes are of the nature of Mars and somewhat of Mercury.

These Are the Stars North of the Ecliptic. The bright ones in the Small Bear are of the nature of Saturn and somewhat of Venus.

Those in the Great Bear, which are *Ayish* and her Children,[1] are [similar in nature to] the co-mixture of the Moon and Venus.

The bright ones in the Dragon are [similar in nature to] the co-mixture of Saturn and Mars.

The stars of the Inflamed One (Cepheus) are [similar in nature to] the co-mixture of Jupiter and Saturn.

Those in the lance of the Armed hero (Bootes) are [similar in nature to] the co-mixture of Saturn and Mercury.

The bright one in the Northern Crown is [similar in nature to] the co-mixture of Venus and Mercury.

The One Walking on His Knees (Hercules) is of the nature of Mercury.

The bright one that is the Falling Eagle (Lyra) is [similar in nature to] the co-mixture of Venus and Mercury.

The bright star called the Hen (Cygnus) and those in the Woman Sitting on a Chair (Cassiopeia) are [similar in nature to] the co-mixture of Venus and Saturn.

Those in the Horseman Carrying the Devil's Head (Perseus) are [similar in nature to] the co-mixture of Saturn and Jupiter. Those on his sword are [similar in nature to] the co-mixture of Mars and Mercury.

The bright one called Ayyuk (Capella) is [similar in nature to] the

[1] *Ayish* was previously identified as the Small Bear, Ursa Minor. The two Bears seem to be mixed up here.

co-mixture of Mars and Mercury.

The stars in the Bearer of the Animal (Ophiuchus) is of the nature of Saturn and somewhat of Venus. Those on his back are [similar in nature to] the co-mixture of Saturn and Mars.

The Flying Eagle (Aquila) is [similar in nature to] the co-mixture of Jupiter and Mars.

The stars of the fish called Dolphin are [similar in nature to] the co-mixture of Saturn and Mars.

The bright ones in the Archer are [similar in nature to] the co-mixture of Mars and Mercury.

The stars in the Woman Who Has No Husband (Andromeda) are of the nature of Venus, and those in the Triangle are of the nature of Mercury.

These Are the Stars South of the Ecliptic. The bright one in the mouth of the Southern Fish is [similar in nature to] the co-mixture of Venus and Mercury.

The one with the Sun[1] is of the nature of Saturn, and the one on the shoulder of the Hero is [similar in nature to] the co-mixture of Mars and Mercury.

The other bright stars[2] are [similar in nature to] the co-mixture of Saturn and Jupiter.

The stars of the River are [similar in nature to] the co-mixture of Saturn and Jupiter.

The stars of the Hare are [similar in nature to] the co-mixture of Saturn and Mars.

The stars around the later Dog (Canis Major) are [similar in nature to] the co-mixture of Venus and Mercury.

Those on the mouth of the Dog, *Al-Sha'are al-Yamania*, (Sirius) are of the nature of Jupiter and somewhat of Mars.

The stars of the Warrior (Orion) [are of] the nature of Saturn and Venus.

Those in the Ship are of the nature of Saturn and Mars.

The bright one called *Ksil* is of the nature of the nature of Jupiter

---

[1] This does not make any sense in this context. Most likely text corruption. [Additional by RH] The Latin has *pectore* 'chest' or 'breast' instead of 'Sun'.

[2] No name specified. [Additional by RH] These all appear to be stars of the South Fish, Piscis Australis.

and Saturn.

The two dim ones are [similar in nature to] the co-mixture of Venus and Mercury.

Those in the image of the Horse[1] are [similar in nature to] the co-mixture of Venus and Jupiter.

The bright ones on the neck are of the nature of Venus and somewhat of Mars.

Those in the Southern Crown are of the nature of Mercury.

The stars of the Hyena (Lepus) are [similar in nature to] the co-mixture of Saturn and Mercury, and the stars of the Censer are of the nature of Jupiter and somewhat of Mercury.

Those are the stars that the ancients have observed.

---

[1] Not sure which constellation. Probably Centaurus.

Chapter Three[1]

The third chapter deals with the aspects of the degrees in the wheel, their friendship and hatred, the quadrants of the wheel and its twelve divisions, and what they indicate at any [given] moment.

The aspect is in four ways, and they are the sextile aspect, the quartile aspect, the trine aspect and the opposition aspect. The signs [forming] the aspects are seven — the 3rd, the 4th, the 5th, the 7th, the 9th, the 10th, and the 11th. The aspect to the 3rd and the 11th is the sextile; the aspect to the 4th and the 10th is the quartile; the aspect to the 5th and the 9th is the trine; and the aspect to the 7th is the opposition.

The aspect to the 3rd, the 4th, and the 5th is northern (sinister) and the aspect to the 9th, the 10th and the 11th is southern (dexter).

The sextile aspect is one-sixth of the wheel which is 60 degrees. The quartile aspect is one quarter of the wheel which is 90 degrees. The trine aspect is one third of the wheel which is 120 degrees. The opposition aspect is one half of the wheel which is 180 degrees. For example, if the Ascendant is at the beginning of Aries, then its northern (sinister) sextile aspect is at the beginning of Gemini and its southern (dexter) sextile aspect is at the beginning of Aquarius; its northern quartile aspect is at the beginning of Cancer and its southern quartile is at the beginning of Capricorn; its northern trine aspect is at the beginning of Leo and its southern trine aspect is at the beginning of Sagittarius. The opposition aspect is at the beginning of Libra.

The strongest of the aspects is the aspect of the opposition. Next in strength is the quartile aspect, then the trine aspect, and the weakest of all is the sextile aspect. The opposition aspect is complete enmity and the quartile aspect is of semi-enmity. The trine aspect is complete friendship and the sextile aspect is half a friendship.

The signs that have no aspect between them nor [are they of the same elemental] nature are four — the 2nd, the 6th, the 8th and the 12th. The weakest among them are the 6th and the 12th. Some of the signs have enmity by aspect yet friendship in another way either because they have the same ascension [time], or same strength, or by ecliptical position.

Those whose ascension [time] is equal are Aries and Pisces, Virgo

---

[1] Heading not in the original. [RH]

and Libra, Taurus and Aquarius, Leo and Scorpio, Gemini and Capricorn, Cancer and Sagittarius.[1]

Those of the same strength are the sign whose crooked [temporal] hours are equal, like Cancer and Gemini, Taurus and Leo, Aries and Virgo, Pisces and Libra, Aquarius and Scorpio, Capricorn and Sagittarius.[2]

A planet in one of the even [temporal hours] signs is called the master (commanding) and the one in the opposite degree in one of the crooked signs is the slave (obeying).[3]

[Aspectual connection between] signs in the ecliptic is true for every two signs that are the domiciles of one planet, such as Aries and Scorpio [that belong] to Mars, Taurus and Libra [that belong] to Venus, Gemini and Virgo [that belong] to Mercury, Sagittarius and Pisces [that belong] to Jupiter, Capricorn and Aquarius [that belong] to Saturn. Even though the Moon has [only] one domicile and the Sun has [only] one domicile, as they are both rulers,[4] their domiciles are considered as belonging to one ruler.[5]

Aries and Libra, Capricorn and Cancer, Virgo and Pisces, and the others, even though they are of the same active nature, there is enmity between them because of the opposition aspect.

The Wheel is divided at every minute of the hours into four parts. The

---

[1] These are the Commanding and Obeying signs of Ptolemy and are also contra-antiscia, i.e., signs whose degrees have their degrees in contra-parallel of each other. See Ptolemy Book I, chapter 15.[RH]

[2] These are the signs that "see" each other and are also called "equipollent" because the length of the daylight when the Sun is either of these signs is the same. See Ptolemy, Book I, chapter 16. [RH]

[3] This paragraph refers to the second paragraph preceding, not the one immediately preceding. [RH]

[4] Possibly meaning the presiding rulers of the day and of the night, as opposed to the other planets which are called attendants or ministers.

[5] This is an interesting concept, which I have not seen elsewhere. In the context of this section it probably means that Cancer and Leo, even though they have no aspect between them, can still be considered as related since they belong to the ruling luminaries. [Additional by RH] This concept of aspects between signs of the same domicile is also found in Paulus (among other places) where they are called *homozōnia,* or 'like-engirding'. See Paulus Alexandrinus, chapter 13.

quadrant of the wheel from the middle of the heaven to the rising degree is oriental, male, and goes forward. Corresponding to it among the four elements is air.[1] Of the [humors in the] human body it is the blood, of the time of the year it is the hot season, of the four parts of the day and the night it is the first one, of the ages of life it is childhood, and of the colors it is the white.

The quadrant extending from the Midheaven to the setting degree is southern, weak like a female, and its actions go backward.[2] Corresponding to it among the four elements is fire, of the seasons it is the summer, and of the four parts of day and night it is the second one. Of the [humors in the] human body it is the red bile,[3] of the ages of life it is young adulthood, and its color is red.

The quadrant extending from the setting degree to the degree of the lower heaven is occidental, male, and goes forward. Of the [four] elements earth corresponds to it, of the seasons it is winter,[4] of the four parts of the day and night it is the third one, of the [humors in the] human body it is the black bile, of the years [of life] it is near 50, and its color is black.

The quadrant extending from the line of the lower heaven to the rising degree is northern, female, and goes backward. Of the [four] elements Water corresponds to it, of the seasons it is the cold, of the four parts of the day and night it is the last one, of the [humors in the] human body it is the cold that mixes with the phlegm, of the time of life it is old age, and its color is green.

The part of the wheel that is above the earth, which is from the

---

[1] This is different from Ptolemy which has the Ascendant of the nature of the Dry and the Midheaven of the nature of the Hot. This would make this quadrant dry moving toward hot. However, this scheme is more consistent with the nature of the elements and the qualities in a cyclical relationship. [RH]

[2] This attribute is not clear, yet obviously intending to mean something negative. [Additional by RH] Here is another possibility. A point in the masculine quadrants is being carried away from the horizon by the diurnal motion. A point in the feminine quadrants is being carried back toward the horizon.

[3] Usually called Yellow Bile. [RH]

[4] This reference to winter along with other such references in the signs suggests that the Hebrew word for winter does not correspond to ours in the order of the seasons but rather corresponds to autumn. The Latin also has autumn here as it does in every other case where the Hebrew has winter. [RH]

ascending degree to the descending one, is called [the] right [side] (dexter, south) and the part below the earth is called left (sinister, north). The two quadrants that are male are also called right and the two female ones are called left.[1] The two quadrants extending from the Midheaven to the Ascendant and from the Ascendant to the Lower Heaven are called the ascending hemisphere, and the other half extending from the Lower Heaven to the Descendant and from the Descendant to the Midheaven is called the descending hemisphere.

At every moment the wheel is divided into twelve parts, corresponding to the number of signs, and they are called houses. [Every] four of them are called by the same name, and these are the 1st house, the 4th, the 7th and the 10th, all four [of which] are called 'Poles',[2] for they are like <points>.[3] The 2nd house, the 5th, the 8th and the 11th, the four of them are called by one name which is 'adjacent to the poles'.[4] The 3rd house, the 6th, the 9th and the 12th, the four of them are called by one name which is 'weak'.[5] The poles are stronger than the adjacent ones, and the adjacent ones are stronger than the weak ones. The strongest of the poles are the 1st and the 10th, the strongest of the adjacent [houses] are the 5th and the 11th, and the strongest of the weak [houses] are the 3rd and the 9th.

At every moment the houses of the wheel correspond to the four elements. I shall give an example from the poles. When the rising sign is Aries, which is of the fire element, at the Midheaven [we find] Capricorn which is of the earth element, the descending sign is Libra which is of the air element, and the lower heaven is Cancer which is of the water element. It is the same for the adjacent and the weak houses.

These twelve houses are fundamental in the nativity of a person, in inquiries, in elections, and also in judgment for the world in general.

The First House is the one ascending from the beginning of the eastern [horizon] line. It denotes life, the body, speech, the mind, fruitfulness

---

[1] This is in fact the way that these quadrants would be seen if one first looked east and noted which quadrants were on the right and left in front of oneself, and then looked west and did the same. [RH]

[2] The Latin predictably has 'angles'. [RH]

[3] Supplied from a variant text.

[4] The Latin again has "succedents to the angles." [RH]

[5] The Latin has "cadents from the angles." [RH]

(fertility), the beginning of all actions, and the thoughts on one's mind. Of a person's age it denotes the beginning. The first ruler of the triplicity[1] indicates life, the character of the newborn, or the inquirer, and his desires, and all that will befall him, whether good or bad, in the beginning of his life. The second ruler of the triplicity indicates the body and [its] strength, and the middle of [one's] life. The third ruler of the triplicity, [which] is the co-ruler, mixes with its [other two] companions and all that they indicate regarding the [[end of]] life.[2]

The Second House denotes money and possessions, and trading, and food, and one's assistants, and those who obey one's command, and the witnesses[3] and the keys and the treasures. The first ruler of the triplicity indicates money in the early years of one's life; the second ruler of the triplicity indicates [it in] the middle years; and the third [ruler indicates

---

[1] The text can be read as 'the ruler of the first triplicity' and continues in the same way for 'the ruler of the second triplicity' and for the 'ruler of the third triplicity'. This does not make sense when discussing one house which normally is identified with the sign on its cusp sign. I changed it to read 'the first ruler of the triplicity' etc., also because that is the known way of assessing signs/houses condition in many other astrological texts. Another reason that can justify this rephrasing is the fact that in the following sections for the other houses the text is not consistent and at time is phrased the way I did it. Therefore, I have kept this phrasing for all twelve houses.

[2] This system in which each triplicity ruler signifies a different attribute of the house is also found in Alchabitius where it is attributed to one *Allendezgod* (in Latinized Arabic). Bonatti also mentions these but has gotten them from the same source as Alchabitius, if in fact his source was not Alchabitius. For each house in this section I will cite the Alchabitius version for the same of comparison. [Translation from the Latin text of Alchabitius by the editor, RH.] "And Allendezgod *On the Nativity* says that the first lord of the triplicity first of all signifies the life and nature of the native or querent, and his pleasure and desire, and what he loves and what he hates, and what of good or bad may come to him in the beginning of his life. And the second lord of the triplicity life and body, and power and strength, and the middle of life. And the third lord of the triplicity signifies what its comrades [the other two triplicity lords] signify and signifies the end of life in death."

[3] The Latin also has "those who testify." [RH]

the same at] the end [of life].[1]

The Third House denotes brothers and sisters, and relatives, and in-laws, and knowledge, and the knowledge of the Torah and the laws, and dreams [interpretation], and modesty, and counsel, and faith, and letters and rumors, and short distance travel. The first ruler of the triplicity signifies the older brothers; the second [ruler signifies] the middle ones; and the third [signifies] the little ones.[2]

The Fourth House denotes the father, land, houses and fields, and regions, and building, and hidden treasures and all hidden things, and the end of any matter. The first ruler of the triplicity signifies the father; the second [ruler signifies] land; and the third [indicates] the end of all matters.[3]

The Fifth House denotes the son (children), gambling,[4] food, drink, [fine] clothing, pleasure, gifts, emissaries, and the crops and the treasures of the father.[5] The first ruler of the triplicity denotes the

---

[1] From Alchabitius: "And Allendezgod spoke about the lords of the house of substance, to wit, of the first, the second, and the third. See which is them is stronger in essence and place. You will make this rightly the source and significator of substance, which, if it is in the Midheaven, he will find substance from the king; and if it is in the house of faith, also of trust, it is better. Similarly the first lord of the triplicity gives substance in the beginning of life, the second in the middle, and the third in its end." [RH]

[2] From Alchabitius: "Allendezgod said that the first lord of the triplicity of the house or brothers signifies elder brothers, the second middle brothers, and the third younger brothers; and their dignity and their state of being will be according to their places." [RH]

[3] From Alchabitius: "Allendezgod said that the first triplicity lord of the house of fathers signifies fathers, the second cities and lands, and the third the ends of things and prisons."

[4] The Hebrew word *mazal* (מזל) means an astrological sign, but also 'luck', and is used in the name for dice games (משחקי מזל), hence 'gambling', which is a known attribute of the 5th house.

[5] As the second house from the fourth.

children, and the ancestors[' property];[1] the second indicates pleasure; and the third signifies emissaries.[2]

The Sixth House indicates chronic illness, slaves and maids, small animals, prison, lies and slander. The first ruler of the triplicity denotes illness and defects; the second signifies the slaves; and the third whether they are useful or harmful.[3]

The Seventh House indicates women,[4] and intercourse, and dispute, and war, and standing trial, and robbers,[5] and partnership, and fighting. The first ruler of the triplicity signifies the women; the second indicates wars; and the third [indicates] the partners.[6]

The Eighth House denotes death, and inheritance, and loans, separation, fear, grief, and loss. The first ruler of the triplicity indicates death; the second [indicates] anything ancient;[7] and the third [indicates] inheritance.[8]

---

[1] The word 'property' has been added because this the second from the fourth which makes this the ancestors' property. But the Hebrew does omit the word 'property'. The Latin omits this reference entirely. [RH]

[2] From Alchabitius: "Allendezgod said that the first lord of the triplicity of the house of children signifies children and life, the second delight, but the third embassies." [RH]

[3] From Alchabitius: "Allendezgod said that the first lord of the triplicity of the house of infirmities signifies infirmities and bad health because of infirmities, from evil circumstances, and injurious wounds; the second lord signifies house servants and servants in general; the third lord signifies what will happen because of these and their usefulness and works, . . ." [RH]

[4] Obviously from a man's point of view. [RH]

[5] This seems a bit out of place for the usual significations of the 7th house, but in Horary astrology this house signifies the person inquired about, which in case of robbery would be the person committing the act.

[6] From Alchabitius: "Allendezgod said that the first lord of the triplicity of the seventh house signifies wives, the second contentions, and the third commingling and sharing." [RH]

[7] This is an unknown attribute of the 8th house.

[8] From Alchabitius: "Allendezgod said that the first lord of the triplicity of the house of death signifies death, the second precepts and old things, and the third, Almaverith, that is, those things which will be inherited from the dead."

The Ninth House denotes travel and a distant road, and everyone who is removed from his high position,[1] and philosophy, and religion and the worship of God, and emissaries,[2] and rumors, and dreams, and oaths, and divination, and laws, and judgment. The first ruler of the triplicity indicates travel; the second [indicates] the faith; and the third indicates wisdom (education).[3]

The Tenth House signifies the mother, government (authority), reputation, and all professions. The first ruler of the triplicity indicates the mother; the second indicates one's rank; and the third indicates one's profession.[4]

The Eleventh House denotes honor and grace, and good name, and hope, and friends and companions, and the king's ministers and treasurers and wardrobe stewards. The first ruler of the triplicity indicates hopeful thoughts; the second indicates friends; and the third

---

[1] This is a very interesting meaning of the 9th house that comes from the fact that it is cadent, falling away from the powerful angle of the 10th house through the diurnal motion about the pole, and therefore it is weak. Thus, when the ruler of the 10th house is in the 9th, it may be interpreted as being removed from power, or may be retiring from a managerial job in order to teach in school or withdraw into religious life.

[2] Here is another non-typical attribution of the 9th house, which is similar to the 'emissaries' of the 5th house, and could also be used for 'diplomats' 'ambassadors' etc. Both the 5th and the 9th are the farthest places from the Ascendant while in a beneficial aspect to it, thus they can represent the native in far away places.

[3] From Alchabitius: "Allendezgod said that the first lord of the triplicity of the house of journeying signifies journeying and everything that happens to one; the second signifies faith and religion; the third is the significator of wisdom, dreams, of stars and auguries, and the truth or untruth of those who practice in these arts."

[4] From Alchabitius: "Allendezgod said that the first lord of the triplicity of the house of kingship signifies work and high rank, to wit, the lifting up of position, and the highest dwelling; the second therefore [signifies] the voice of command and boldness in the same; the third signifies the stability and durability of this." [RH]

whether they are good to him or bad.[1]

The Twelfth House denotes grief, poverty, jealousy, hatred, fear, fraud, vigilance, prison, captivity, all disgrace and affliction, animals [that serve] for riding [or pulling vehicles]. The first ruler of the triplicity indicates grief; the second [indicates] prison; and the third [indicates] enemies.[2]

---

[1] From Alchabitius: "Allendezgod said that the first lord of the triplicity of the house of trust signifies trust, the second friends, and the third the usefulness and profit of these." [RH]

[2] From Alchabitius: "Allendezgod said that the first lord of the triplicity of the house of enemies signifies enemies, the second workers, and the third beasts and cattle." [RH]

Chapter Four[1]

The fourth chapter [deals with] the nature of the seven planets, their influence, and all that they indicate. I shall begin to mention Saturn as he is the most elevated of them, since het is in the 7th sphere from the Earth.

Saturn is cold and dry and his nature is evil and harmful. He indicates destruction, and ruin, and death, and grief, and mourning, and weeping and crying, and ancient things. In his share of the human spirit is the power of thought. His is the first region of the Earth,[2] which is India. Of the nations, in his share are the black people, the Jews, the Berbers, and in general all old people, and farmers, and builders, and tanners, and those who clean lavatories, slaves, the inferior, robbers, ditch and grave diggers, and undertakers.

In his share of the metals of the earth are black lead and rusty iron, and black stones, and those that are blue and black, and the magnet stone, and every heavy black rock.

In his share [of places] of the earth are the caves, and wells, and pits, and prisons, and every dark and uninhabited place, and the cemeteries.

Of the animals [in his share are] the elephants, the camels, and every animal that is big and ugly, such as hogs, wolves, monkeys, black dogs and black cats. In his share of the birds is every large bird with a long neck, like the ostrich, the eagle and the vulture, and every bird that has a strange voice, and the crow and the bat and every bird of black appearance. Of the things that crawl he [rules] the fleas, bugs, flies, mice, and all harmful, foul, creeping creatures inside the earth.

In his share of the trees are the tree-with-gall, the carob tree, the one called acorn, the bramble, and every tree that has thorns that harm and bears no fruit, and lentils and millet.[3]

Of the medicinal [plants he has] the cactus called aloe, and '*al-hilag*', and '*al-blilag*', and '*al-malag*', which are similar to the '*prunash*' which come from India, and every plant that is toxic, and everything bitter such as '*la'ana*', and in general every black plant.

---

[1] Heading not in the original text. [RH]

[2] The climata again. See page 15, note 6. [RH]

[3] The Latin has *panicum*. [RH]

His nature is cold and dry, and his flavor is the astringent and everything that has disagreeable taste and foul smell.

Of the spices he has the *'al-kasat'* which is the *'kidda'*, and *'kashur al-azar'* which are [from] the bark of a tree, and '*aslika*' and '*al-ma'a*' which is gum.

Of garments he signifies the cloak, woolen clothing, and blankets, and all heavy garments.

In his share of the human nature is the contemplation, little talk, astuteness, isolation from people, controlling them and winning, and levying taxes, and getting angry, keeping one's word, deep thought, knowledge of secrets and the worship of God, inferior people, contrariness, fear, worry, coveting, and in general habitually lying with little usefulness and much ruin.

[Saturn indicates] tilling the land, building, mining metals, seeking hidden treasures, digging and examining things of the dead, and everything that lasts for many years.

His occupation is every work that is exhausting and of little reward, and all inferior work, such as quarrying stones, cleaning ditches, and all filthy work.

He indicates fathers and grandfathers, the deceased, crying and separation, wandering, poverty, humiliation, distant bad roads where danger lurks, and he has no success in all undertaking.

If he rules the nativity, he will bestow the good things he has, provided that he is on the favorable side of the Sun and in [an advantageous] division of the wheel and of the moment, as I shall explain; and if it is the opposite, he will yield every detested thing.

When [Saturn is] in an elevated degree and mixing with a good planet, all that he indicates will turn to be beneficial if the benefic planet is in its high elevation, and if it is in its low position, its usefulness will be little.[1] If a malefic planet is mixing [with Saturn] and both are in their power,[2] then it indicates the greatest victory and pillage. If the malefic planet is not in its power,[3] it will produce all that is vulgar and inferior with disgrace and shame. All this I shall explain

[1] It is not clear here whether this is a reference only to elevation on the deferent, or whether epicyclic elevation is counted as well. Later authorities tend to combine them as both being factors in elevation. See page 22, note 1.

[2] That is, they are both essentially dignified. [RH]

[3] That is, outside of its essential dignities. [RH]

in the *Book of Nativities.*[1]

If Saturn [rules] alone in the figure of a man, his stature will be erect, his appearance in the middle between white and dark; he is modest, and noble, and strong; his hair is black and curly, with hair on his chest, and his eyes are medium [in size] with a lot of cold and moisture. All this is if Saturn is oriental[2] of the Sun, as I shall explain. If Saturn is occidental, he will be thin, his hair is not curly, and his nature is cold and dry. In general it indicates one who is ugly in appearance and looks, with much hair on his body, his nostrils are thick and so are his lips and his teeth, and his smell is foul.

In his share of the human body are the bones, the spleen, the right

---

[1] Ibn Ezra's *Book of Nativities* is planned as part of this series. [RH]

[2] The terms 'oriental' and 'occidental' have been causes of much difficulty in the history of astrology. I will not attempt to deal with all of that difficulty here, but we do need to call attention to what is the most peculiar aspect of the the use of these words. When the a planet is oriental and near the Sun, it is not *east* of the Sun. It is *west* of the Sun! Instead the *Sun* is *east* of the planet. The problem is the same with 'occidental' where the Sun is west of the planet, not the planet west of the Sun. (See Joseph Crane, *A Practical Guide to Traditional Astrology,* chapter 6, an ARHAT publication.) The problem seems to have arisen because in several ancient languages, notably Greek and and Latin, the words for 'east' and 'west' have their origin in the words for 'rising' and 'setting' respectively, as in "the Sun rises in the east and sets in the west." Consequently a confusion has come into existence between the two sets of meanings of the words. While there is little difficulty in regard to sunrise and sunset, the problem comes with heliacal risings and settings where the same words are used. But for a planet to rise ahead of the Sun in the morning it must be in advance of the Sun, i.e., west of the Sun. Likewise, for a planet to set in the west after the Sun at sunset, it must be backward from the Sun, i.e., east of the Sun. In other languages such as Hebrew the confusion becomes more acute because authors such as Ibn Ezra have used the Hebrew words for 'east' and 'west' where we would use oriental and occidental which are the Latin-derived words for 'rising' (heliacal rising) and 'setting' (heliacal setting).

In English language texts as well as in Latin texts this problem is complicated by the fact the phrases such as "oriental of the Sun" or "occidental of the Sun" are used when the phrases really should be "rising with respect to the Sun" or "setting with respect to the Sun." We have used the phrase "oriental of the Sun" and "occidental of the Sun" in this text following custom, but the reader should be aware of the true meanings of these phrases. [RH]

ear, the place of the urine (bladder), and the red bile.[1] His ailments are the insanity, senility, tremor, the plague and the sickness of hemorrhage, leprosy, sickness of the legs, every chronic illness, and those of cold and dryness that come from coldness and dryness.

In his share of human life is old age and the end [of life], of the [geographical] directions it is the east, of colors it is the black and the color of dust, of the days it is Saturday, and of the nights it is the night [before] Wednesday [day], of the [planetary] hours it is the first and the eighth; his letters are the *shin, ayin* and *peh* (ש, ע, פ),[2] and some say that the *nun* (נ) too. Of the glyphs it is both of these [*see glyphs at right*].[3] His greatest years are 256, his great years are 56, the mean years are $43^{1}/_{2}$, the lesser years are 30, and the years of the part called fardar[4] are 11. The influence of his body is 9 degrees before and also after him.[5]

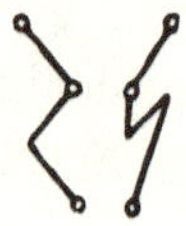

Jupiter is below Saturn for he is in the sixth sphere from the Earth. He is warm and moist and temperate in his nature. He is the best of the planets, indicating life and increase of beneficence and fruitfulness and multiplication and justice and honesty. In his share is the vegetative soul. Of the regions in his share is the second region[6] which is the land of Yemen or *Sheba*, and the land of *al-Haziz* that is Assyria, and *Meisha*. The nations [under his rule] are the Persians, and the Babylonians, and of people it is the judges, the scholars, those who serve God, the modest, the pious, the generous and the righteous. In his share of the earth minerals are tin which called '*tutia*', the stone called '*yakut*', the white, the saffron, sapphire, onyx, crystal, and any white

---

[1] Usually Saturn is assigned to black bile. [RH]

[2] The Hebrew letters in parentheses are written right to left following normal Hebrew practice while their names in English letters are written from left to right. This is done throughout the remainder of the text. The reader should be aware of this. [RH]

[3] The glyphs shown here for Saturn and for the other planets are actually the ones contained in the manuscript on which this translation is based. The reader will note that they are not very similar to the ones in use at present. [RH]

[4] Also known as firdar, or alfridary. [RH]

[5] This is the orb of light of Saturn. [RH]

[6] See page 15, note 6 on the climata. [RH]

and brilliant stone that has usefulness. In his share are the houses of prayer and worship of God, and the clean places. Of the animals he has those that have cleft hoof such as sheep and deer, and any pleasant looking animal that does not harm but little. Of the birds [he has] the peacock, chicken and doves, and in general any bird that eats seed and is useful to people. In his share of [creatures that crawl] on the ground is everything that is useful and not harmful such as the silk worm. In his share of the trees are the walnut, the almond, the fruit called pistachio, *'bandak'* and pine nut, and in general any [fruit] where the outer shell is removed and the inside is eaten. In his share are wheat, barley, peas, rice, and of flowers it is the one called *'al-bahar'*, jasmine, *'al-marangush'*, and the like. Of the medicinals it is everything of good mixture that has good fragrance and good flavor, such as musk and *'al-kapur'* or *'kapra'*, sugar, *'al-basbasah'* which is called *'al-ud'* and amber and [others] similar to them.

Of garments it is the fine [kind] such as cotton clothes and any delicate garment.

In his share of human nature is esteem, justice, peace, faith, modesty, good name, generosity of the heart, freedom of the soul, speaking the truth, keeping [one's] covenant, smiling face, love of good, abhorring evil, and in general hate of anything not in accordance with the law. He loves to talk a lot and desires praise[1] from people. Most of his thought is directed towards seeking wealth and accumulating it, and winning in everything, in an honest way though. [He seeks] knowledge of the laws and jurisprudence, and solving dreams, and service in synagogues.

[Jupiter] also signifies children and grandchildren, and in general he indicates good fortune, grace and honor. When he is (rules) alone in the nativity[2] of a person, he brings by nature every honorable thing, and the opposite if [Jupiter] is not in his [own] power.[3] The [other] planets increase or decrease [his influence] when they mix with him whether by conjunction or aspect. If the stars are malefic, they decrease his goodness, and if they are benefic, they increase his goodness.

---

[1] This could also be 'thankfulness', which is not too much out of the Jupiterian character either.

[2] The Hebrew word is *mitboded* (מתבודד), which means 'solitary', 'alone', 'keeping to oneself'.

[3] Meaning 'not in any essential dignity'.

As for the figure of the individual, if [Jupiter] is oriental, he indicates an attractive stature, a body of white and reddish appearance, little hair, beautiful eyes, and he is splendid in all his activities, and his (Jupiter's) nature is warm and moist. When he is occidental, one's body will be erect, but its whiteness will not be clean, and his hair is the opposite of curly (straight), and he is bald, and moisture increases in him.

In general he indicates a person who has a kind soul, benevolent nature, pleasant appearance, respectable looking beard and thin hair.

In his share of the human body are the liver, the left ear, the ribs and the blood vessels.[1] Of ailments it is every one that passes quickly. Of a person's years [it rules] the period between adulthood and old age; of the colors [it rules] the white, the green, the saffron, the even,[2] and every pretty and bright color. Of the [geographical] directions [he denotes] the north, of the days [he rules] Thursday, of the nights [he rules] the night [before] Monday,[3] and of their [planetary] hours [he rules] the first and the eighth. Of the letters he rules *het*, *samekh* and *bet* (ח, ס, ב). His symbol is [*see glyphs at right*]. His greatest years are 427, the greater years are 97, the mean years are 45, and the minor years are 12. The years of the part [called] Al-fardar are 12, and the [orb of] influence of his body is 9 degrees before and after.

Mars is hot, dry, burning, harmful and destructive. He indicates ruin, drought, fires, rebellion, bloodshed, killing, war, fighting, blows, separations, and in general everything that is not right as it should be. Of the human nature he denotes anger.

Of the Earth in his share is the third region[4] such as Egypt and

---

[1] The text is *kol ha'dam* (כל הדם) which translates as 'all the blood'. If we add one letter, which may have been dropped by a copyist, then we get *klei ha'dam* (כלי הדם) meaning 'blood vessels'. [Additional by RH] The Latin has "the whole of the blood."

[2] Not sure what is meant here. The word is *hayashar* (הישר), which normally means 'the even' or 'the straight', or, in other contexts 'the honest' or 'upright'.

[3] In the Jewish tradition the night comes before the day, therefore, in this expression this is the night before Monday.

[4] See page 15, note 6 on the climata. [RH]

Alexandria. Of the nations [he rules over] the people of England and fire worshippers. Of people in his share are all men of war and those who command it, robbers, ironsmiths, blacksmiths, makers of lances and swords, bloodletters and attendants of animals, and the like.

Of the earth's minerals [he rules] iron, red copper, sulfur, and naphtha. [He rules] glass vessels, and all weapons, and every red stone. In his share are fortifications, towers, and the *'purni'*. Of the animals [he rules] hyenas, dogs, and the tiger; of birds [he rules] *'al-buzia'* or *'astur'*; of crawling creatures [he rules] everything that is harmful such as the snake and scorpion, and everything that has deadly venom; of plants [he rules] the brier, and *'agsis'* called *'al-bacam'* which is a kind of wood used for dyeing; also *'al-azfur'*, and *'al-fuah'*, and every tree that has thorns, and pepper, mustard, cumin, radish, leek, garlic and *'al-sadav'* or *'roda'*. Of medicinal herbs [he rules] *'astucdus'*, *'tarbad'*, *'sababing'*, *'zarion'* water, *'povion askemonia'*, *'parsion'*, *'carcac'*, and every plant that causes immediate pain, and everything whose flavor is hot and sharp, and unpleasant. Of the aromas [he rules] *'al-sandal'* (sandalwood?) and saffron; of clothing [he rules] rabbit-skin and *'al-samur'*, and garments dyed in *'al-carmaz'*.

In his share of human nature is speed, courage, victory, strength, fighting, dispute, pillage, anger, insults and curses, lies, informing [against others], debauchery, rudeness, not keeping one's promise, making a false vow, permitting all that is forbidden by the Torah, stealing, every wicked deed, cruelty, much effort, wandering from place to place, getting in harm's way, torture of people and beating, imprisoning and taking captive, robbing of money, digging through walls and opening doors to discover any hidden thing.

In general he is all evil with nothing good in him. He signifies brothers,[1] and quarrelsome people, and abortions, and a woman's miscarriage, and any bad accident that happens suddenly and every calamity.

If he [rules] alone in the nativity, he will give the individual the good [qualities] that he has provided that he is in his power and aspected by benefic stars. It is the opposite when he is not in his power

---

[1] Mars' signification of 'brothers' comes from the fact that it is in third sphere from Saturn (counting down), and third place corresponds to third house. This is also mentioned in his *Book of Reasons*. [Additional by RH] This attribute of Mars is also found in Hindu astrology.

and [is] mixed with malefic stars. When he is oriental of the Sun, the native will be tall with white appearance tending towards ruddiness, with much hair on his body. When he is occidental [of the Sun], he (the native) will be short, his appearance will be reddish, his face round, his eyes small, his hair red and not curly and the nature of his body is dry.

In general he indicates anyone whose appearance is red, his eyes are like those of a cat; he is ugly in looks, he has blemishes and pockmarks on his face, and he acts with haste, quarrel and mania.

In his share of the human body are the gallbladder, the right nostril, the tendons, the kidneys, and the penis. [Also] in his share are the red bile, and the burning blood.[1] Of its ailments are ague fever, tuberculosis, red pox that break out on the body, anxiety, insanity, bruises and burns.

Of human life [he rules] the period of adulthood. His appearance is very red. Of the [geographical] directions [he rules] the west; of the days [he rules] Tuesday, and of the nights [he rules] the night [before] Saturday [day], of their [planetary] hours [he rules] the first and the eighth. Of the letters of the alphabet [he has] *tsade*, *qof* and yod (צ, ק, י); of the symbols he has [*see upper glyph*]. And some say [*see lower glyph*]. His greatest years are 284, his greater years are 66, his mean years are $40^{1}/_{2}$ and the minor years are 15. The part called Al-fardar is 7 years, and the [orb of] influence of his body is 8 degrees before and after.

The Sun is very hot and dry, and [can be both] useful and harmful, doing good and bad. Hers[2] is the air and the sensitive[3] soul. In its share of the regions is the fourth,[4] such as the land of Babylonia, Iraq and the land of Israel; of the nations [it rules over] the Edomites, the Turks, and

---

[1] Mars traditionally rules the blood of the arteries as opposed to the blood of the veins. Before the discoveries of Harvey in the 17th Century these were regarded as being two different fluids. The phrase 'burning blood' refers to the brighter color and the fact the blood contains more oxygen although the ancients were not necessarily aware of that. [RH]

[2] A reminder that the Sun is usually but not always feminine in Hebrew. Therefore, this section the English will go back and forth between feminine 'she' and the neuter 'it'. [RH]

[3] The word can also be understood as 'feeling' or 'sensing'.

[4] See page 15, note 6 on the climata. [RH]

'Dilam'; of people [it signifies] kings, ministers, and counselors; of the minerals of the earth [it rules] gold and precious stones such as *'al-as';* [also] in its share are red salt, marcasite which looks like gold, *'al-sharnaa'*, and any sparkling stone. In its share of [places in] the land are palaces and the houses of royalty.

Of living things [it rules] humans, horses, lions, and the large rams; of the birds [it rules] *'al-ukvan'* or *'lammergeier',* and *'shudanik'*; of the crawling things [it rules] the large ones that kill; of the plants [it rules] date trees, vines, olive trees, apple trees, mulberry trees, *'hav al-alukh'* or cherry, *'spargalim'*, and fig trees; in its share is the rose, *'aluvia'* and silk; of the aromas [it rules] *'shibolet'*, 'nerd', *'al-carmaz'*, lilac, and any kind whose characteristic is hot and spicy in taste.

In its share is every dignified quality. Of human nature it has knowledge, intelligence, majesty, beauty, courage, seeking high positions, desire for wealth, garrulity, expeditiousness, and excess of desire. Of trades [it signifies] gold and silver smiths, and crafting of crowns.

It also indicates constitutions and laws, and the union of society, and fathers and the middle brothers.[1]

According to its strength it will deliver of its nature. When it [rules] alone in the nativity, the individual will be fat with white appearance; his eyes are medium [size], his hair is thin, and he is dignified looking. In its share of the human body are the heart, the right eye by day and the reverse by night, the brain,[2] the arteries and [in general] the whole right half of the body. It also indicates red bile and of its ailments is anything that occurs in the mouth.[3]

Of [a person's] life its period is [young] adulthood; its color is pure red and saffron; its flavor is spicy; of the [geographical] directions [it denotes] the east; of the letters of the alphabet [it has] *alef*, *dalet* and *lamed* (א, ד, ל),

[1] It is not clear where this attribution comes from. Is it because the Sun is below Mars and Mars signifies brothers, as was explained before?

[2] The brain is ruled by the Moon! Check the *Book of Reasons*!

[3] This seems to be a scribal error, based on other references. In allocating the parts of the face the mouth belongs to Mercury. [Additional by RH] Schoener in the *Opusculum Astrologicum* also gives ailments of the mouth to the Sun, Book II, Canon 2.

and this is the symbol [*see glyph above*];[1] of the days [it rules] Sunday, and of the nights [it rules] the night [before] Thursday [day], and of their [planetary] hours [it rules] the first and the eighth. Its greatest years are 1461, the greater years are 120, the mean years are 39½, and the minor years are 19. The years of the Al-fardar are 10, and the [orb of] influence of its body is 15 degrees before and after.

Venus is cold and moist, of favorable and even nature. She is a good planet that denotes the desiring [part of the] soul, fructification and multiplication.

In her share of the regions are the fifth[2] where Spain is, and some of Idumea; of the nations [she rules over] the Arabs and all those of the Islamic faith; of people [she signifies] young boys, eunuchs, women, actors, musicians and poets. In her share is everything that is in the belly of the earth, the saffron [colored] copper, *'al-lazurar'*, magnesium, *'al-martakh'*, *anu'shadar'*, *'al-zag'* or *'irmant'*, *'atincar'*, and all ornaments and women's rings.

In her share of [places in] the land are gardens, orchards, places of myrtle and flowers, places of women (harems), and festivities, and beds. In her share of the animals are deer, and gazelles, and any beautiful animal; of the birds [she rules] *'al-higgil'* or partridge, doves, and birds; of the crawling creatures [she rules] the spider and ants. Frogs are also in her share. Of plants [she rules] apples, pomegranates, and any fruit that has good fragrance and pleasant taste. In her share is [also] balsam and anything that has good aroma and is succulent and oleaginous.

Of clothing [she denotes] all embroidery and pretty garments; of human nature [she denotes] cleanliness, friendship, laughter, playfulness, merriment, dance, pleasant talk, love, adultery, playing dice, generosity, excessive lust for everything, false testimony, tendency towards drunkenness, natural and unnatural sexual intercourse, love of children, fondness for the marketplace,[3] and in general love of justice and of places of worship.[4] Of the trades [she denotes working with] anything

---

[1] This glyph is identical to the standard Byzantine glyph for the Sun except that it is rotated 180°. [RH]

[2] See page 15, note 6 on the climata. [RH]

[3] Loves to shop?

[4] These last two attributes seem out of place here, for Venus as we know her. Transposed from Jupiter?

that is dyed, and in sewing. She indicates all eating and drinking, the mother, daughters, and the younger sister.

According to her strength in the nativity her nature will be manifest in the person. When she [rules] alone in the nativity and oriental, the person will be fat, his appearance will be white, his shape pretty, his eyes black, and he is tall. When she is occidental, the person will be short, the whiteness of his face will not be pure, his hair will not be curly and he will be bald. In general she indicates anyone who is good looking with round face, black eyes, and he is smiling.

In her share of the human body is the flesh, the milk, the liver,[1] and the semen. All moisture [in the body] belongs to her. Of her ailments are all that occur in the kidneys and the genitalia.

Of the period of life [she rules] adolescence, as a rule after age 13. Her colors are white and tending somewhat towards green. Her flavors are sweet and bland. Of the [geographical] directions [she rules] the east; of the letters of the alphabet [she has] *tet* and *dalet* (ט, ד). Her symbol is [*see glyph at right*]. Of the days [she rules] Friday, and of the nights [she rules] the night [before] Tuesday [day], and of their [planetary] hours [she rules] the first and the eighth. Her greatest years are 1151, her greater years are 82, her mean years are 45, and the minor years are 8. The years of the part called Al-fardar are 8. The [orb of] influence of her body is 7 degrees before and after.

Mercury is mixed and variable since he changes according to the nature of the rest of the stars and the signs, and his nature is somewhat tending towards coldness and dryness. In his share are the soul of man and the power of intelligence.

In his share of the regions is the sixth one,[2] and of the nations it is *'Gog and Magog'* and the people of India. Of people [he signifies] the philosophers, scientists, doctors, scribes, mathematicians, geometers, merchants, artisans of etching and painting.

In his share of the metals are live silver (mercury) and all coins and etched stones. In his share are the marketplaces, schools, all workshops,

---

[1] The liver is usually given to Jupiter. Another copyist error? [Additional by RH] Schoener's *Opusculum Astrologicum* also gives the liver to Venus along with Jupiter.

[2] See page 15, note 6 on the climata. [RH]

springs, rivers and fountainheads.[1] [Of living things,] in his share are humans, foxes, wild donkeys, and any animal that is light in jumping; of the birds [he rules] the starling. In his share are the bees and any bird that takes flight nimbly; of the crawling creatures [he rules] worms; of trees [he rules] the citrus, and partly the pomegranate, walnut, reeds, cotton-wool, and flax. In his share are various kinds of resin, ginger, *'gandava destar'* which are the testicles of an animal found in England whose tail is like that of a fish, *'aakar karkha'*, *'madg'*, *'asrukh'*, *'pakakh'*, *'al-dacar'*, *'al-zarnabar'*, *'al-shitrag'*, *'al-gintiba'*, and anything that is sour in taste. Of clothing [he denotes] linen garments and all those which have embroidery in them.

In his share of human nature are speech, intelligence, education, wisdom, science of the stars, divination, all sorts of magic, eloquence, accuracy of language, fast talk, ability to recite and rhyme, knowledge of hidden secrets and prophecy; [also] kindness and compassion[2] and avoiding evil, [as well as] musical talent, love of anything miniature, negotiating, verbal arguments without resorting to blows, all sorts of trickery and deception, and writing forged documents. His hand is trained in all skills, and he is eager to do all [kinds of] activity, and [he is inclined towards] acquiring wealth and squandering it. His nature varies according to the individual nativity. He also signifies the younger brothers, and his nature will manifest according to his strength in the sign.[3]

When he [rules] alone in the nativity and oriental, [the person] will be short, with a small head and beautiful eyes. When he is occidental, one's appearance will be the average between black and white, he will be thin, his eyes will be small and dryness will predominate in him. In general he denotes a person who has a prominent forehead, long

---

[1] The last three attributes seem out of place here, since Mercury is usually not associated with water.

[2] This word is a close guess only. The Hebrew word *yakar* (יקר) does not properly mean respect, but 'dear', both as in 'affection' and as in 'expensive'. However, in literary language it appears as an adjective that describes a person who is respected in society and is estimable. [Additional by RH] The Latin has *charitas* from which comes our word charity but which actually means a chaste, non-erotic love.

[3] Another translation, based on a footnote is: ". . . according to its strength in the nativity."

nostrils, with little flesh, thin beard, long fingers, and he is a person of intelligence and learning.

In his share of the human body is the tongue, the mouth, and the nerves, and he partakes in the [rulership of the] blood. Of his ailments are those of the soul such as thinking and worry and doubt. Of human lifetime he governs childhood. His color is blue and all shades of colors that are mixed together. His flavor is sour, and his [geographical] direction is north. Of the letters of the alphabet [he has] *vav*, *tav*, *resh*, *mem* (ו, ת, ר, מ) and letters of continuation. This is the symbol [*see glyph at right*] Of the days [he rules] Wednesday, and of the nights [he rules] the night [before] Sunday [day], and of their hours the first and the eighth.

His greatest years are 480, the greater years are 76, the mean years are 48, the minor years are 20, and the part called Al-fardar is 13, and the influence of his body is 7 degrees before and after.

The Moon is cold and moist but there is some heat in her.[1] She corrupts all bodies and putrefies them. When she is opposite the Sun, she will ripen all fruits, and in her share is the power of productivity.[2] In her share of the regions is the seventh,[3] which is the limit of the inhabitable world, of the nations it is *'Al-Tsvia'*, and of people it is sailors, travelers, messengers, and servants. In her share of the metals are silver, crystal, onyx, and lime. In her share are the seas, the rivers, and the ponds. Of the animals [in her share are] mules, donkeys, cattle, the hare, and fish; of the birds [in her share is] any white bird, and of crawling creatures any white [colored] one. Of the trees [in her share are] the river willows, peach trees, and all sorts of vegetables, [such as] squash, and watermelons, and cucumbers; [also] in her share are *'al-karpa'* or cinnamon, and *'dar'* peppers, and *'dar sini'*, and every species whose characteristic is cold and moist, and her flavor is salty, and her color is white with some green. In her share of the fragrances is the *'al-saudi'*, and also in her share are the clothing and the spreads.

In her share of human traits are much contemplation and expression

---

[1] As it nears a Full Moon it gets illumination from the Sun.

[2] The phrase is *ko'akh hatoledet* (כח התולדת) which can also be understood as 'the strength of the nature [of the chart]', or may be even 'giving birth'.

[3] See page 15, note 6 on the climata. [RH]

of emotion, stupidity, forgetfulness, timidity, naiveté, revealing secrets, fondness of gaiety and festivity, knowledge of magic, storytelling, [telling] lies, calumny, and much eating. In general she tends to vary according to each nature [of sign or planet she is with]. She governs the years of infancy, the mother and her sisters, and the older sisters, and women, and any pregnant woman.

Her nature will manifest according to her strength in the nativity. When [she rules] alone in the nativity of a man, his posture will be erect, his complexion will be white with some yellow, his face will be round, and his eyebrows are like stuck together, and he is swift in his motion and his gait. From the beginning of the month until the middle she indicates whiteness, and from the middle [of the month] to the end she indicates some blackness.

In her share of the human body by day is the left eye and the reverse by night, the lung, the throat, the stomach, the womb, and the left side of the whole body. Of her ailments is every illness that comes from increase of moisture. Of a person's years [she governs] infancy during the period of suckling. Of colors [she indicates] the gray and the green, of the flavors the salty, of the [geographical] directions the right of west, of the letters of the alphabet *zayin* and *ayin* (ז, ע). Of the symbols [she has] [*see glyph at right*], of the days [she rules] Monday, of the nights [she rules] the night [before] Friday, and of their hours the first and the eighth. Her greatest years are 520, the greater years are 108, the mean years are 39½ [1], the minor years are 25, the years of the part called Al-fardar are 9, and the influence of her body is 12 degrees before and after.

[1] Should be 69½. See comment in *B.O.R.*

Chapter Five[1]

The fifth chapter deals with the good [conditions] of the planets and their harmful [ones], their power and their weakness.

[Positions of strength:]

When a planet is in a conjunction with a benefic planet, or in aspect with it [by] sextile, quartile, or trine, and not aspected by malefic planets nor [are they] conjunct with it, or when a planet separates from a benefic planet and conjoins a[nother] benefic planet, or when a planet is in the middle between two benefic planets, or in a conjunction with the Sun, or in aspect with it, a trine or sextile, or in aspect with the Moon and the Moon is with the benefic planets, or when a planet is swift in motion, or increasing in light and therefore separating,[2] or when a planet is in one of its places of dominion such as its domicile, exaltation, triplicity, bound, face, or in its place of elevation,[3] or in swift degrees,[4] or received,[5] or in its similitude,[6] or when the two luminaries are in a place ruled by benefic planets, then it is as if they are in their own dominion, and likewise it is for the benefic planets when [found] in a place ruled by other benefic planets.

When a planet is in one of the positions I have mentioned, then its nature is strengthened and it will indicate every good thing according to its [essential] nature. If it is the opposite of all the ways that I have mentioned, then its nature weakens. If two or three or more of the positive configurations I have mentioned coincide [for the planet], then any goodness that it indicates will increase and will be stronger.

The goodness[7] of a planet comes in three ways. One is double goodness, the second is whole goodness, and the third is medium goodness. Double goodness is when two or three dignities combine such

---

[1] Heading not in original. [RH]

[2] From a conjunction with the Sun.

[3] The auge. See page 22, note 1.

[4] This may be reference to a certain category of degrees (unknown to me), or simply saying that the planet is direct in motion past a station, where it slows down. [Additional by RH] The Latin here has 'bright degree'.

[5] By an aspecting planet. See page 124.

[6] See page 125.

[7] The word is *tova* (טובה) meaning 'goodness' or 'benefit'.

as when Mercury is in the sign of Virgo; then it has two dignities, one being in its own domicile and the other being in its exaltation; and if Mercury is [also] in its bound, that would be three dignities.[1] The [one] whole goodness is when the planet is in its [one] domicile that is [more] adjusted[2] to its nature, [such as] Saturn in Aquarius, Jupiter in Sagittarius, Mars in Scorpio,[3] Venus in Taurus, Mercury in Virgo, and the luminaries in their [respective] domiciles. The medium dignity is when a planet is in its domicile that is not adjusted to its nature: Saturn in Capricorn, Jupiter in Pisces, Mars in Aries, Venus in Libra, and Mercury in Gemini.

[Other types of] strength for a planet is when it is ascending in the northern side (latitude), or when it ascends in the wheel of elevation and descent[4] whose solid [sphere] is distant from the solid [sphere] of the Earth, or when it is in its second station where it begins to go direct, or when it is coming from under the light of the Sun, or when it is in one of the poles or their adjacent [houses], or when the three superior planets (which are Saturn, Jupiter, and Mars) are oriental of the Sun and

---

[1] Mercury's bound in Virgo is degrees 1 - 7. Mercury is the only planet that has both domicile and exaltation in the same sign, Virgo, and so sweeps three dignities in one easy step.

[2] The Hebrew word is *tityasher* (תתישר) for which a literal translation would be 'become straight or direct or even with', but the following text implies adjustment or conformity with. [Additional by RH] This is the doctrine of joys or thrones in which each planet (except for the Sun and Moon) is stated to have more rejoicing in one of its domiciles than the other. This arises from the fact that every one of the starry planets has more dignity in one of its signs that the other. Thus: Saturn — Aquarius (domicile and triplicity), Jupiter — Sagittarius (domicile and triplicity), Mars — Scorpio (domicile and triplicity), Venus — Taurus (domicile and triplicity), Mercury — Virgo (domicile and exaltation). The reader will also note that the "joy" by sign of a planet is also the domicile in which agrees in sect with it. This is found in Dorotheus and many medieval sources. Later medieval and renaissance sources misunderstood this doctrine and changed it.

[3] For Mars' nature by sect and sex see page 10 for Ibn Ezra's unconventional but very interesting position on the sex of Mars. [RH]

[4] Certainly this refers to the motion to and from the auge on the deferent, but may also refer to epicyclic elevation as well. See page 22, note 1.

in sextile aspect with her, and thus are in good condition,[1] or when they are in one of the male quadrants or male signs, and the same is true for the Sun except when it is in Libra.[2]

The strength of the three inferiors (which are Venus, Mercury and the Moon) is [manifest] when they are occidental of the Sun or in the female quadrants.

The weakness of a planet is when it slows down in its motion, or when it is in its first station, or moving backwards. The retrograde motion is more detrimental for the two inferiors (Mercury and Venus) for they get burnt by the Sun or the planet is under the light of the Sun; or [when a planet is] in the dark degrees, or in the opposite of its similitude, or in its house of fall, or declines towards the south, or is southern, or in one of the falling (cadent) houses, or in the Fiery Road which is from 19 Libra to 3 Scorpio,[3] or when it in its house of detriment, or when it is with a retrograde planet, or when it is in its house of fall, or in a cadent house, or when it is not received, or when a planet is like an outsider in its place.[4] It is even worse when [the ruler of the place] does not aspect it, or when the three superior planets are occidental of the Sun, or when [they are] in one of the female quadrants or female signs. This is [also] true for the Sun except when she is in the 9th house which is her house of joy.[5]

---

[1] When the superior planets are oriental of the Sun, they are also in their proper sect, a phase that begins at their conjunction with the Sun until the opposition when they become occidental and go out of their proper sect. While they are in their proper sect, they also become retrograde, around the point when the faster moving Sun forms a trine with them. If this reasoning is correct, then the dexter sextile of the superior planets to the Sun is the true benefic position within the sect! Mars is out of line again, as its inclusion in this category violates its other classification as a nocturnal sect planet.

[2] Because Libra is the fall of the Sun.

[3] Note that the Fiery Road is defined from the fall degree of the Sun to the fall degree of the Moon. [Additional by RH] At this point we have no way of knowing whether this is meaningful or coincidence.

[4] Peregrine — having no rulership of any type in that place.

[5] The 9th house is in a female quadrant. [Additional by RH] The ninth house was originally called "the house of the Sun-god." Even though it is in the feminine quadrant, it is the joy by house of the Sun.

[Positions of weakness:]

The weakness of the inferior planets is when they are at the beginning of their oriental [position] of the Sun or when in one of the male quadrants.

The harm of a planet is when it is in a conjunction with malefic planets, or opposite them, or in a quartile, trine or sextile aspect [with them] and the distance between it and the malefic planet is less than the limit of the planet;[1] or when the planet is in the bound of a malefic or its domicile; or when one of the malefics is more elevated than the planet as when in the 10th house or the 11th house relative to the place of the planet. It is worse when the malefic does not receive the planet; or when the malefics are in a conjunction with the Sun or a quartile aspect or an opposition; or when a planet is with its south node or its north node, or with the Dragon's Head or with its Tail with less than 12 degrees between them. It is most difficult for the Moon when [found] in this situation [with the nodes]. For the Sun it is difficult when the distance [to either node] is 4 degrees before or after her. The ancients said that the nature of the Dragon's Head is to increase and the nature of the Tail is to decrease so when the benefic planets are with the Head, it will increase their good, and when the malefics are with it, it will increase their evil; [in the same way] when the benefics are with the Tail, it will decrease their good, and when the malefics are with it, it will decrease their evil. That is why the Hindu astrologer says that the Head is good with the benefics and bad with the malefics.[2] The weakness of a planet is [also] when it is between two malefics in a conjunction or by aspect, as I shall explain later.

The Harm of the Moon [Comes] in Eleven Ways:

One, when she is eclipsed. Two, when she is under the light of the Sun with less than 12 degrees between them whether she is approaching him or separating from him. Three, when there is the same number of degrees [of orb] when the Moon is opposite the Sun whether before the opposition or after it. Four, when she is with the malefics whether in a

---

[1] The orb of influence? Not clear.

[2] This doctrine is widely found cited in medieval renaissance astrology text, but it does not seem to be widely observed. This is one of the few, if not the only, source that attributes this doctrine to the Hindus. I cannot find it in any of the sources of Hindu astrology available to me. [RH]

conjunction or by aspect. Five, when she is with[in] the influence of [the] 12[th division belonging to] Saturn or Mars.[1] Six, when she is with the Head or the Tail [of the Dragon] with less than 12 degrees between them. Seven, when she is [in the] southern [latitude] or declining [towards it].[2] Eight, when she is in the Fiery Road.[3] Nine, when she is at the end of the sign for those are the bounds of the malefics. Ten, when she slows down in her motion. This you can know when her motion in one day is less than her mean motion, which is written in the tables. Eleven is when she is in the 9th house because it is opposite her house of joy.[4]

---

[1] The Hebrew reads literally as follows: "Five, when she is with the influence of 12 of Saturn or Mars." [Additional by RH] We do not know if the "twelfth" referred to here is a one-twelfth division of the signs ruled by the malefics or the signs themselves. A similar ambiguity occurs in Greek texts especially in Ptolemy. In any case the logic would be the same; the Moon is not benefited by being in subdivisions ruled by malefics.

[2] The meaning of this is a little unclear. Does he mean all southern latitude or only moving southerly in latitude and perhaps being at the southernmost latitude? [RH]

[3] The *via combusta.* [RH]

[4] The Moon's joy is the third house which was known to the Greeks as the "House of the Moon Goddess," or simply as "Goddess." [RH]

Chapter Six[1]

The sixth chapter describes the [circumstances of the] planets in their own [motion] and their [circumstances with] regard to the Sun. The circumstances of the planets are in many ways.

A planet may be in the middle of its wheel, which means that there are 90 degrees between the planet and the beginning [position] of its elevation equally to the left and the right. Or, a planet may be ascending in the wheel of elevation or descending in it, or at the beginning of the elevation, or descending from the middle of the wheel towards its nadir, or ascending from its nadir to the middle of the wheel, or at its nadir.

A planet may be in increasing motion and thus [increasing] its light and the strength of its body, or diminishing in all these, or in a mean position, neither increasing nor decreasing; [it can also be] increasing or decreasing in number, or its calculation may be increasing, decreasing, or [of] mean [value]; or its latitude may be northern going up or down, or southern going up or down, or it may have much latitude or little latitude or no latitude [at all]. When a planet is 90 degrees away from the beginning of its elevation, then it is direct in its motion and the reason is that it is [in an] intermediate [position], and so is its light and the strength of its body.[2]

When a planet is less than 90 degrees away from the beginning of its elevation and moving towards it, then it is ascending [in the wheel] and decreases in its motion, and light, and strength of body.

When it is in its place of elevation, then the decrease is complete; and when it descends from the its elevation towards [the point of] its half-wheel, then it increases in motion, and light, and strength of body;

---

[1] Heading not in original. [RH]

[2] There is a difficulty with this passage. All other references to 'elevation' refer to the movement of the planet on its deferent rather than epicycle. But here we have the Hebrew word *yashar* (ישר) which has a variety of meanings such as 'direct', 'upright', 'straight', etc. Either this is a rare reference to the planet's motion in its epicycle or *yashar* means something quite different here such as 'proper' indicating an *average* motion. The Latin does not help much as it differs from the Hebrew very widely, but it does not say anything about being "direct". So this is either a rare (in this text) reference to epicyclic motion, the text is corrupt, or *yashar* has an unusual meaning in this instance. [RH]

When it descends from that half-wheel towards its nadir, it also increases in all [its attributes]. When it is in its nadir, then it increases in all of these completely.[1]

The same things that are said for the Moon are also said for the superior planets, for when she is 12 degrees away from the Sun, it is said that her light is increasing until she comes to the opposition of the Sun, and from there until she joins the Sun again her light wanes.

If you wish to know when a planet increases in number, then look when you enter its number in the columns of the direct degrees [in the tables]. If it is [in the] first [column], then it is increasing, and if [in] the second, then it is decreasing, and if the number is not [found] there, then it is neither increasing nor decreasing. If you add the calculation to its mean position, then it is considered increasing in number, and if you subtract the calculation, then it is decreasing in number. If you neither add nor subtract, then the planet is in [its] orbit.[2]

For Venus, when you subtract the calculation of the Sun from the mean motion and there is no remainder, or 180 [degrees] remain, then it is with the Sun in the same degree.

For the superior planets, if their motion in one day is greater than their mean motion, then it (the motion) is fast in motion, and if less, then it is slow.

For Venus and Mercury, look to see if their motion is less than the mean motion of the Sun, then they are slow in their motion; and if more, then they are fast; and if [their motion is] the same as the mean motion of the Sun, [they are] neither slow nor fast.

When the latitude of a planet is northern and that is when it is between its [own] north node until 90° [from that position], then it is [called] northern-ascending,[3] and from there to the tail (its south node)

---

[1] For a complete explanation of all of this see the notes to pages 14-17 of our edition of Abu Ma'shar's *The Abbreviation of the Introduction to Astrology.*

[2] The planet is on its deferent circle. Again see the notes to pages 14-18 in our edition of Abu Ma'shar's *Abbreviation of the Introduction to Astrology.* [RH]

[3] The prose is a bit strange here, but the Latin confirms the Hebrew. There is some kind of textual corruption or linguistic ambiguity in the original. It should read something like this, "When the latitude of a planet is northern and when it is between its [own] north node until 90° [from that position], then it is [called] northern-ascending . . ." [RH]

it is northern-descending. From the tail to [the next] 90° it is southern-descending, and from the 90° [position back] to the north node it is southern-ascending. When [the planet] is 90° equally distant from the north and the south node, it has its maximum latitude;[1] as it gets closer to the head or the tail its latitude decreases, and when it is with the head or the tail, it is on the ecliptic.

The circumstances of the planets with regard to the Sun [occur] in many ways. From the time when the three superiors conjoin the Sun in the same degree until they are in opposition to the Sun they are right (dexter) of her, and from the point of opposition until they join with her [again] they are left (sinister) of her.

Only Venus and Mercury: from the time when they separate from the Sun, [and] retrograde until they become direct in motion [again] and follow her and overtake and join with her, then they are dexter of her. From the time they separate from the Sun, [being] direct in motion and becoming occidental, until their station and [then] going back while the Sun overtakes them and they join with her, then they are left of her.

Likewise, the Moon from the moment she separates from the Sun until she comes to her opposition is left of her (the Sun), and from the opposition to the time of the conjunction she is right [of the Sun].

Know That the Three Superior Planets have variations [in their regard] to the Sun in sixteen ways.

One, when the planet is with the Sun in the same degree and minute, then the planet is called joined provided that it is away from the Sun by less than 17 minutes; for if it is more than that, it is [called] combust; but when it is joined, it indicates everything good in all affairs [that it indicates].[2]

If the planet is further than the mentioned [17] minutes up to 6 degrees[3] and it is oriental, then it is combust, and that is so for Saturn

[1] North or south. [RH]

[2] Cazimi. Note that Ibn Ezra does not make any mention of whether or not the planet is also within 17' minutes of arc in latitude. Many other authorities require that. [RH]

[3] Eight degrees is a more common figure. [RH]

or Jupiter but Mars is called combust up to 10 degrees.[1]

When they are [distant] more than 6 degrees, or Mars is more than 10 degrees, up to 15 degrees, they are said to be under the light[2] [of the Sun], and this is the third way. When a planet is combust, it has no power at all. When it comes out from the combust zone and it's under the light, then its power returns to it somewhat, and the further away the better. The ancients said that Mars should be 18 degrees away and then it comes out from under the light. Then they will have their average strength in all their indications.

They are called oriental and strong [from that point] until they are distant from the Sun by a sextile aspect, which is when the planet is in its greatest strength, and that is the fourth way.

From that point until they are distant from the Sun by a quartile aspect their power of brilliancy will diminish,[3] and this is the fifth way.

From there to its first station its power continues to weaken, and this is the sixth way.

When the planet is in its first station, then its strength is completely weakened; and this is the seventh way.

When it is going retrograde towards the opposition of the Sun, it is in the eighth way and it has no power at all.

When a planet is opposite the Sun, it is the ninth way; and that indicates confusion, and hope which does not materialize.

From the time they separate from the opposition to the Sun is the tenth way, which is when it gains some strength until [it comes to] its second station.

When it is there (the second station), which is the eleventh way, then its power returns to it.

From there, until it is distant from the Sun by 90° and occidental, then its power diminishes somewhat, and that is the twelfth way.

From there it continues to weaken until it is away from the Sun by

---

[1] Later authors usually give the same arc of combustion for all planets. This reference may be a survival of the tradition of giving each planet its own arc of combustion. See Julius Firmicus Maternus, *Matheseos,* Book II, chapter ix. There the following values are given: Saturn 15°, Jupiter 12°, Mars 8°, Vanus 8°, Mercury 18°. [RH]

[2] Known to us as "under the beams" or "under the rays." Other sources give 17° or 18°. [RH]

[3] As they slow down approaching the first retrograde station.

a sextile aspect, and that is the thirteenth way when only one third of its strength remains.

When it is 15 degrees away from the Sun, that is the fourteenth way, and then none of its power is left.

When the planet is under the light of the Sun, it is the fifteenth way; and when [it is back] in the combust zone, it is the sixteenth way.

[But] Venus and Mercury are [evaluated] in a different way. [From the time] when they are [in conjunction] with the Sun in [the same degree and] minute, or in less than the [17] minutes mentioned above, and they are oriental, [they are] in two ways for they are combust until their distance from the Sun is 7 degrees. [And from that point] until they are away 12 degrees away from the Sun they are under the light, and from there their strength continues to grow, and they are [still] oriental, until they are in the first station, and that is the fifth way. Then their strength diminishes until they get near the Sun [again]; and [when] the distance between them is 12 degrees, they are under the light; and when the distance is 7 degrees, then it is [in] the combust zone until they are with the Sun. When they are separating from the Sun and occidental, they are in the combust zone until 7 degrees, and from there until 12 degrees they are under the light, and when they come out from under the light, their strength continues to increase as long as they are direct in motion until they begin their station and turn retrograde; then their strength is removed until they are close to the Sun by less than 15 degrees where they under the light, and then [they proceed into] the combustion [zone].

The Moon has sixteen ways with regard to the Sun.

One, when she is joined with it before the Sun or after it with[in] 16 minutes, and this is the first way.

The second way is when she is 6 degrees away [from the Sun] and she is occidental,[1] and then her strength begins somewhat.

The third is when she is 12 degrees away, and from there her strength increases until her distance is 45 degrees, when her [first] quarter is illuminated.

And from there until 90 degrees her strength still continues to increase and half of her is illuminated.

---

[1] At the New Moon.

From there until there are 135 degrees between her and the Sun, three quarters of her are illuminated.

From there until her distance from the opposition to the Sun is 12 degrees she is in her power, and at the moment of the opposition she is in her great[est] strength.

When she moves away from the opposition by 12 degrees, it is the tenth way. [From there] until she is 45 degrees away from the opposition then her light decreases by one quarter.

From there until she 90 degrees away from the opposition then half the light remains in her body.

From that point until there are 45 degrees between her and the Sun one quarter of her light remains.

From there [she proceeds] until the distance is 12 degrees, where she enters the [area] under the light, and from 6 degrees [she is] under the combustion [zone].[1]

Know that when you leave out the two times when she is under the light and the two times when she is under the combustion [zone], then twelve ways remain. These are called the Keys of the Moon which are greatly needed [in order] to know the matters of rains.

---

[1] In this whole section it is interesting to notice that the basic scheme is the same as the one put forth in modern times by Dane Rudhyar of dividing lunar cycle into 45° segments. The main difference is the finer division resulting from the 6° and 12° before and after the New and Full Moon, and of course the Moon in Cazimi. [RH]

Chapter Seven[1]

The seventh chapter deals with the circumstances of the planets which are thirty:

1. Application (הקרוב)
2. Conjunction (החבור)
3. Co-mixture (הממסך)
4. Aspect (המבט)
5. Separation (הפרוד)
6. Solitary Motion (void of course) (הלוך בדד)
7. Feral (השומם)
8. Transfer [of Light] (העתקה)
9. Collection [of Light] (הקבוץ)
10. Return of Light (השבת האור)
11. Conferring of Influence (תת הכח)
12. Conferring of Rulership (תת הממשלה)
13. Conferring of Nature (תת התולדת)
14. Conferring of Two Natures (תת שתי התולדות)
15. Directness (היושר)
16. Distortion (העזות)
17. Prevention (המניעה)
18. Returning of Good [Influence] (השבה לטוב)
19. Returning of Harm[ful Influence] (השבה לרע)
20. Cancellation (refranation) (הבטול)
21. The case [of Three Planets in One Sign] (המקרה)
22. Loss (frustration) (האבוד)
23. Deprivation of Light (כריתות האור)
24. Pleasantness (Recovery) (הנועם)
25. Recompense (התגמול)
26. Reception (הקבול)
27. Generosity (הנדיבות)
28. Similitude (הדמיון)
29. Besiegement (האמצעיות)
30. Authority (השררה)

[1] Heading not in the original text. [RH]

Application is when two planets are in one sign, direct in motion, while the degrees of the lighter planet are fewer than those of the heavier planet, and as long as its degrees are less it is called *applying,* and when it [joins] with the [heavier] planet in the same [degree and] minute, then the application is complete. The beginning of the application is when there are less than 15 degrees between them, and the fewer the degrees [between them] the stronger is everything that they indicate. In the same way is the application to the rays of light which are the seven aspects as we mentioned [above].

Conjunction is when there are two planets in one sign and each one is in the [orb of] influence of the body of the other [planet], then they are said to be joined. And when they are in same [degree and] minute then, their power is complete in everything that they indicate. If one [planet] is in the [orb of] influence of another body, yet the other [body] is not within the [orb of] influence of the first one, then whatever they indicate will be but half complete. An example is when the distance between the Moon and Saturn is 8 degrees, before or after, then each one of them is within the [orb of] influence of the other. But when there are 10 degrees between them, then Saturn is within the influence of the Moon, but the Moon is not within the influence of Saturn.

As long as the lighter planet approaches the conjunction with the heavier one it will have more power than from the time of the separation [from the conjunction] even though it is [still then] within the influence [of the weightier one].

If the two planets are [found] in two separate signs, yet each one is within the influence of the other, they are not said to be joined because they are in different signs, and that is the opinion of the ancients. But, I, Abraham, the scribe of this book, differ with them, as I shall explain in the *Book of Nativities.*[1] Nor did the ancients mention the measure of the bodily influence of the superior [fixed] stars which are in the wheel of the zodiac. Only Doronius[2] alone said that it is one fourth of a sign, but the opinion of those who specialize in this study is that it is the measure of bodily influence of the planet, whether the

[1] Ibn Ezra is one of the first authors to unambiguously approve of out-of-sign aspects. [RH]

[2] Dorotheus of Sidon.

superior star is of the first magnitude or the second one.[1]

The explanation for a conjunction is when two or more planets conjoin, and one's wheel is lower than the other's, and they are both in the same degree and minute vis-a-vis the ecliptic, the lighter one will occult the heavier one for the observing eye for it (the lighter) goes underneath it (the heavier), and this occurs if there is no [difference in] latitude from the ecliptic between them as I shall discuss later.

Co-mixture is when a planet joins with another and they are of the attendants;[2] then from their [two] natures another nature is produced, such as Saturn and Mars both of which are malefic; and when they join, the ancients said that it indicates good but the truth is that one cancels the action of the other, so the native is saved from harm. Therefore, they do not indicate good but the good [that comes] from not being harmful.

When Jupiter is joined with Saturn (which is the Great Conjunction[3] since they are both superior), its manifestation will be according to the strength of whichever one [is stronger in that place]. When Mars is conjoined with Venus, their nature is temperate, as I shall explain. The conjunction of the planets with the Sun harms them, and the most detrimental is for Venus and the Moon. Saturn and Mars also harm the Sun when conjoined with it,[4] while Jupiter and Venus will benefit and will not harm [it]. Mercury, because of its many motions and because it is always near the Sun, is little harmed when under the light of the Sun or in the combustion zone. When a planet is joined with Sun (according to the ancients), it has great power,[5] thus, they said that when it is so for Mercury, it is as if there are two Mercuries in the wheel; but Ptolemy disagrees with them and he is right. The conjunction

---

[1] The Levy and Cantera translation has all of this previous section concerning only the superior planets. The Latin confirms the interpretation given here which also make much more sense. [RH]

[2] Excluding the luminaries.

[3] When the medieval astrologers referred to "Great Conjunctions," they referred to Jupiter-Saturn conjunctions. [RH]

[4] The Sun is feminine in the Hebrew language but we have translated 'her' as 'it' here. [RH]

[5] This is probably a reference to Cazimi again, i.e., the exact conjunction. [RH]

of the Moon with Saturn and Mars is harmful. If the conjunction with Saturn is when the Moon is decreasing in light, it is worse; and if her light is increasing, it lessens the damage.[1] It is the opposite for [a conjunction with] Mars; and when the Moon is in her [own] power, they (the malefics) will lessen her strength but not harm her much.

The Aspect. The beginning of the aspects, whether it is the sextile aspect, or the quartile, or the trine, or the opposition (as I have explained in Chapter Three), is when it is 12 degrees away from the [partile] aspect, but Ptolemy said 6 degrees,[2] and that is the truth. Let me give an example: When there are 54 degrees between one planet and another, it is the force of the sextile aspect, and when the distance between them is 60 [degrees], then it is in complete aspect and will indicate the perfection of any matter that it signifies. And so it is for the rest of the aspects.[3]

Separation is when the lighter planet passes the heavier planet by 1 degree,[4] whether by conjunction or aspect, and this occurs in two ways. If that lighter separating planet joins another planet by conjunction or one of the other aspects, then its mixture will be with the other planet, but if it does not [proceed to] join another planet, nor aspect it, and there is between it and the [former] heavy [planet] less than its body's orb of influence, it is still within the its co-mixture.[5] More [degrees] than that will cancel the co-mixture.

---

[1] This may be a survival of a teaching found in Julius Firmicus Maternus among others that the waxing Moon makes a more favorable joining with a diurnal planet and, when waning, a less favorable one. For nocturnal planets it is the other way round. See the next sentence. [RH]

[2] We have no idea where this teaching is in Ptolemy. [RH]

[3] This seems to be at odds with his discussions of aspect orbs elsewhere. See under *Conjunction* above. Of course it could be that orbs of influence apply only to bodily conjunctions and not the aspects. But that would be very atypical of the astrology of this period. [RH]

[4] What is not clear is whether Ibn Ezra means that aspect has separated by sixty minutes or that one of the two planets is no longer in the degree of the aspect. [RH]

[5] Now this is an interesting bit. Separating aspects have more impact according to this text if no other application has begun on the part of the swifter planet. [RH]

You always have to look at degrees that are of the same ascension. An example is when one planet is at 10 degrees of Aries and the other is at 20 degrees of Pisces, so the distance of both of them from the Line of Justice[1] is equal and they are considered as being in conjunction. Another way is when they are in the degree that has the same crooked [temporal] hours, such as 17 degrees of Leo and 13[2] degrees of Taurus.[3] The reason is that their distance from the beginning of Cancer is equal. It is the same when their distance [is equal] from the beginning of Capricorn. Those whose distance from the beginning of Aries or Libra are equal are called a straight conjunction, and those whose distance is equal from the northern point [or] the southern one, which are the beginning of Cancer or Capricorn, are called the beholding[4] conjunction.

Solitary Motion (Void of Course) is when a planet separates from another, whether in a conjunction by 15 degrees, or in aspect by 6 [degrees], and does not join another planet as long as it is in that same sign, or no planet beholds it by a complete aspect, whichever it may be.

Feral is when a planet is in a sign and is not aspected by a[ny other] planet as long as it is there, and it is not separating from [any] planet. This happens only to the Moon because of its swift motion.

Transfer [of Light occurs] in two ways. One is when a lighter planet separates from a heavier one and joins with another planet or aspects it. Then it transfers the power of the first heavier planet to the latter.

The second way is when a light planet joins one heavier than itself,

---

[1] Equinoctial axis. [Additional by RH] This is the contra-antiscion.

[2] The text has 14 (י"ד), which probably is a corruption of (י"ג) 13, as it should be, in order to reflect the 17 Leo.

[3] The antiscion. [RH]

[4] Signs of equal distance from the Aries-Libra axis have the same ascensional time, and are traditionally called 'commanding' and 'obeying', the summer signs being the commanding ones. Signs of equal distance from Cancer-Capricorn axis have the same length of day-night, and are called 'beholding' (each other).

Ibn Ezra's phrase is *mah'beret nohakh* (מחברת נוכח), which translates as 'an opposition conjunction'. Levy-Cantera translated it as 'diametrical conjunction'. The word נוכח also implies seeing or becoming aware of, so I chose to translate it as 'beholding', using the traditionally known term.

then the middle one transfers the light of the lighter planet to the latter heavy one.

Collection [of Light] is when two or more planets join with one planet, so that heavier planet takes it all.

Return of the Light is in two ways. One is when two planets do not conjoin nor aspect one another, but both conjoin or aspect another planet , and that other planet aspects the house in question or the planet that is sought so that [third] planet returns the light to the [matter that is] sought.

The second way is when the ruler of the rising sign does not conjoin the ruler of the matter that is sought nor aspects it, or when they separate from each other, then, if another [third] planet transfers the light between them, it is also considered as if it is a conjunction.[1]

Conferring of Influence[2] is when a planet is in its domicile, or exaltation, or triplicity, or bound, or face, and it is joining another planet or aspecting it, then it confers its power upon it.

Conferring of Rulership[3] is when one planet beholds another in aspect of complete friendship (trine) or half a friendship (sextile); then their co-mixture is even (harmonious).

Conferring of Nature[4] is when the planet joins with the ruler of the sign it is in, or the ruler of its exaltation, or bound, or triplicity, or face; then that [ruler] will confer its nature upon the planet.

---

[1] Meaning the matter is perfected as certainly as if through a conjunction.

[2] This is Abu Ma'shar's "Pushing power." See *The Abbreviation of the Introduction to Astrology*, Chapter 3, [31]. [RH]

[3] This *may* be Abu Ma'shar's "Pushing management" but it appears to be a bit garbled. See *The Abbreviation of the Introduction to Astrology*, Chapter 3, [34]. [RH]

[4] This is Abu Ma'shar's "Pushing nature." See *The Abbreviation of the Introduction to Astrology*, Chapter 3, [30]. [RH]

Conferring of Two Natures[1] is in two ways. [One is] when a planet is in a sign where it has dominion and it joins with another planet that has dominion in the [same] sign, or aspects it, such as Venus with Jupiter in Pisces.

The second way is when a planet joins with another one which is of the same nature, diurnal with diurnal, and nocturnal with nocturnal.

Directness[2] is when a planet is in its maximum strength, and that is when it is in one of the poles (angular houses) or the adjacent [ones] (succedent).

Distortion[3] is when a planet is in one of the cadent houses.

Prohibition is in two ways. [One is] when three planets are in one sign and their degrees are unequal so that the heavier one has more degrees. The one in the middle will prevent the lighter planet with the less degrees from joining the heavy one until it (the middle one) will pass, such as when Saturn is at 20 Aries, Mercury at 15 and Venus at 11, so Mercury will prevent Venus from joining with Saturn until it (Mercury) passes [Saturn] and then Venus will join with it (Saturn).

The second is by way of aspect when there are two planets in one sign, the lighter one joining the heavier one, and another planet beholds the heavier planet by any aspect. Then the one in the sign (the lighter planet) will prevent the beholder and destroy its action, provided that their degrees are equal. If the degrees of the beholder are closer to [exactitude of] the aspect, then the joining planet cannot prevent the beholding one.

Returning of Good [Influence]. Know that returning is as follows. When a planet joins with another planet which is under the light of the Sun or aspects [it], but that one cannot receive (accept) the light because of its weakness, so it is returned. Another way is [an aspect] with a retrograde planet, so it returns whatever it received.

The returning of good is in three ways. One is when the one

---

[1] This is Abu Ma'shar's "Pushing two natures." See *The Abbreviation of the Introduction to Astrology*, Chapter 3, [32]. [RH]

[2] 'Advance' in Abu Ma'shar. [RH]

[3] 'Retreat' in Abu Ma'shar. [RH]

holding the influence that is returned to it [also] receives the giver;[1] the second is when the lighter planet is direct in motion and the heavier one is combust, or under the light or, retrograde; and the third is when the returning planet is in one of the falling houses[2] and the planet to whom the light returns is in one of the poles or the adjacent [houses].

Returning of Harm[ful Infuence] is the opposite of the [above] mentioned three ways.

Cancellation (Refranation) is when the planet is within the [orb of] influence of another's. Yet, before it joins it in the same [degree and] minute it turns retrograde.

The Case of Three Planets in One Sign[3] is when one is light and its degrees are many; the second is heavier than him and its degrees are few; and the third is lighter than the first one and it joins the heavy one. [But] the first one with the greater degrees turns back and joins the heavy one.

Loss (Frustration) is when [a planet is] about to join another planet in one sign; but before it is joined, the heavier moves out of the sign and there another planet joins with it.

Deprivation of Light is in three ways. One is when a light planet joins one heavier than itself and both are in the same sign, and that heavy one joins another planet heavier than itself; and before the [first] light one joins with the heavy one, the planet in the next sign retrogrades and enters the sign where the heavy planet is and joins with it, thus cutting off the light from the first planet.

The second way is when a light planet joins with a heavier one both in the same sign, and the heavy planet joins a [yet] heavier one; but before the first reaches the second, [that second planet joins] with the third and passes it, so that it cuts off the light from the first.

The third way is when a planet joins another planet that is not

---

[1] The text is not quite clear as to who is the giver and who is the receiver.

[2] Cadent in the modern sense [RH]

[3] This is Abu Ma'shar's "Resistance." See *The Abbreviation of the Introduction to Astrology*, Chapter 3, [44]. [RH]

needed [for the matter sought after].

Pleasantness (Recovery)[1] is when a planet is in its pit or its inferior position,[2] and another planet is joining it, or it joins another planet, and that planet is its friend or the lord of its house, or it has [some other] dominion in that sign; then it will get it out of its pit or inferiority and comfort it.

Recompense is when that first planet which extricated the one out of its pit or debility is aspected by it, and [now] the extricating one fell into the pit or is in its debility; then that [other now rescued planet] will get it out.

Reception is when a planet joins, whether by conjunction or aspect, another planet which is the lord of its house, or the lord of its exaltation, or the lord of its triplicity, or bound, or the lord of its face, then it is received by that planet. Or, when a planet joins a[nother] planet, and the second planet is in a house of the applying planet or its house of exaltation, then this is reception too. Only, if it is in its triplicity house, or bound, or face, it will not receive it by complete reception unless the two rulerships coincide, the triplicity with the bound or face. The trine aspect and sextile is reception too, also if they are in the degrees of signs that are of equal ascension. A benefic planet receives a[nother] benefic planet because its nature is straight (similar). Mars and Saturn receive one another when they are in conjunction or in sextile aspect or trine, but not in the other aspects.

Reception may be strong, medium, or weak. The reception which is always strong is of the Moon with the Sun because she will receive him from all the signs for her light is from him. Except when the [reception] is from the opposition, it will be with much effort and trouble. If it is in a sign where she has dominion, then it is double reception, like Mercury when receiving another planet from Virgo, for it is [both] its domicile and exaltation; and that is complete reception.

---

[1] 'Favor' in Abu Ma'shar. [RH]

[2] It is not clear from the Hebrew what kind of "inferior position" this is. It looks like opposition to the auge, but other texts suggest that he refers to the planet's fall in the sense of opposition to the exaltation. On the other hand it may be a generic term for debility. [RH]

Reception by domicile is medium, and reception by triplicity, bound or face is weak.

Generosity[1] is when two planets are in each other's domicile, or exaltation, or some other rulership; even though they do not join nor aspect one another, there is [still] reception between them.

Similitude is when a male planet is by day above the earth, in a male sign and in male degrees, and a female planet is by night above the earth and by day under it, in female signs and in female degrees. If it is opposite, then everything that the planet indicates is not proper.

Middle Position is when a planet separates from one malefic planet and joins a[nother] malefic planet by conjunction or by aspect, or when a benefic or malefic planet is in the sign preceding it and another planet is in the following sign. If the Sun aspects the middle planet between the two malefics, it mitigates the harm considerably.

Authority.[2] If the three superior planets are oriental[3] of the Sun, from the time they begin to be visible, then they are in their highest authority until there is a sextile between them and the Sun. From there until the quartile aspect their authority decreases, and from there until the second station they have no authority. If these planets are oriental of the Sun and occidental of the Moon, there is no higher authority than that. [As for] the three inferior planets, their authority begins from the time they are visible in the west after sunset. The power of authority of Venus and Mercury remains until they turn retrograde. And when they are occidental of the Sun and oriental of the Moon there is no higher

---

[1] The heading for this paragraph is *nedavot* or *nedivut* (נדבות), which is translated as 'generosity'. [Additional by RH] This is the modern definition of a mutual reception. In most medieval texts mutual reception requires that the two planets *be in aspect* as well as being in each other's dignity of the same kind. Here and in the Latin version of Abu Ma'shar's *Abbreviation* we find this form of mutual reception without requiring the aspect as a separate kind of mutual reception in both cases called "Generosity." This might be a terminology worth restoring.

[2] This can also be understood as 'strength' or 'power'.

[3] See page 93, note 2.

authority than that. The authority of the Moon is until the middle of the month.[1]

Ptolemy said that if the proportion of the sign the planet is in to the sign where the Sun or the Moon are is in the same proportion as their [zodiacal] domiciles are in their [proper] sectors, then the planet has great strength.[2]

---

[1] The Jewish lunar month, i.e., until Full Moon.

[2] This is the condition known as "proper face" or Almugea. [RH]

Chapter Eight[1]

The eighth chapter [contains] all the judgments for the nativity, for revolutions and for inquiries, and they are one hundred and twenty.

1) As the Moon is close to the Earth and swift in motion, she co-mixes more [frequently] with the planets as she confers her power upon them, but no planet confers power upon her; thus she transfers one's light to the other. She is like a newborn, for her light increases little by little and grows until it is full, and then it decreases little by little until it cannot be seen and it disappears from the world. Therefore, all the ancients said that she indicates every matter, and the thought and the beginning of every activity; and if she is in her power and her circumstances are good, then any undertaking that one begins at that hour will succeed. It is the opposite if she is with the malefics. It is said that you should seek the inquirer [of the question] from the rising sign and its ruler, and the thing inquired about from the seventh sign and its ruler, and always give the Moon co-rulership with it.[2]

2) If the Moon is solitary (void of course), then it indicates anything that is futile, and anything that the inquirer is asking about will not come to pass.[3]

3) The planet which the Moon joins or aspects indicates every future thing and everything that the inquirer hopes for. If the planet is benefic, then [it will turn out to be] good, and if malefic, then [it will be] bad.

4) The separation of the Moon [from a planet] indicates things that have passed. Therefore, if she is separating from a conjunction or an aspect to a benefic planet, then it was good, and if from a malefic, then [it was] bad.

5) If the planet to which the Moon is applying her power is in its

---

[1] Heading not in the original text. [RH]

[2] Whatever this text may mean, the normal practice is to make the Moon a co-significator of the querent not the quesited. [RH]

[3] This is a much more harsh notion of the void-of-course effect than others before and after Ibn Ezra have held. See Masha'allah *On Reception.* [RH]

strength, then the matter will be done properly, and the opposite of this if [that planet] is weak.

6) The same is true for every planet that confers its power upon another one as the thing [sought after] will be [accomplished] according to the strength of the receiver.

7) On a day when the Moon is with the malefics anything a person inquires about will not be properly perfected. If a planet[1] is in the rising sign and the Moon is in one of the poles, it is worse; for this indicates fear in the mind and sickness in the body; and if she is in the cadent houses, this indicates fear but not sickness.

8) The planets are of two types, benefic and malefic, so [for] every place where you find a benefic planet pronounce good, and the opposite if it is the reverse.

9) The benefic planets are right and advantageous whether they are received or not; and if they are received, all the better. The malefics destroy by nature, but if one is received by a [good] planet, it will relieve its evil. It is the same when it aspects [the malefic planet] by sextile or trine.

10) A planet is not declared harmed until the [orb of] light of the malefic is affecting it according to the orb [of light] of its body.[2] If less[3] than that, it indicates only little harm. After separating from it by aspect, even one degree or more, it will cause fear which is not realized. It is the same way for the benefic planet and the planet that signifies the thing [sought after]. If it does not aspect the rising sign, the [thing] wished for will not come to pass.

11) When the benefic planets aspect the malefic ones, they will diminish

---

[1] The main text has the word *kohav* (כוכב) which is star or planet, but a footnote says that one of the manuscripts (used for this edition) has *kohav hama* (כוכב חמה), which makes more sense.

[2] The Latin has the following: "A planet is not described as harmed until the rays of a malefic reach it according to the force of its body." [RH]

[3] Meaning greater distance resulting in the harm being less.

their harm.

12) The benefic [planets] always indicate good and the malefics indicate evil. Only when it (the malefic) is in its great power, then it indicates good, except that it comes with effort and suffering.

13) When the benefics are in [a place] opposite to their nature, or in their houses of detriment, or in their houses of fall, or in cadent houses that have no aspect between them and the rising sign, then they do not bring benefit at all.

14) The malefics too, when in the same situation, they harm but little.

15) If the benefic is in its strength and it is alone in the rulership of the nativity, then its benefit is great.

16) When Jupiter aspects the malefic, it changes its nature into benefit, yet Venus cannot change the nature of Saturn, except with the help of Jupiter. Therefore, Jupiter removes the harm of Saturn, and Venus removes the harm of Mars [even] more than Jupiter [does].

17) When the benefics and the malefics are in a bad place or combust, they indicate something inferior, and they do neither good nor evil because of their weakness.

18) When a malefic is oriental of the Sun, in its power, in a sign where it has dominion, and is not aspected by another malefic, it is better than a benefic planet that is combust or retrograde.

19) When the malefics are the significators of the thing inquired about and the ruler of the rising sign is joined with them, and the Moon [is aspecting] from a quartile aspect or opposition, if the thing desired comes to pass, it will turn out bad.

20) When the malefic is in the rising sign and it has some dominion there, this will relieve its harm; and if it is retrograde, this will heap evil upon evil.

21) If the malefic beholds the benefic by a quartile aspect or an

opposition, it decreases from its[1] goodness.

22) If the malefic is in its place of rulership and in one of the poles or the adjacent [succedent houses], its influence is considered as [that of] the benefic planets.

23) If the malefic is a stranger (peregrine) in its place, its evil will increase.

24) When the malefic is in one of the poles, or beholds a planet from the quartile aspect or opposition, then its harm is complete, even if it (the aspected planet) is stronger than the [malefic] planet. [If the aspect is] from the trine or sextile aspect, the harm will be alleviated.

25) If the malefic is the significator of the thing inquired about, it indicates delay, and it will not come except with misery and worry.

26) If a malefic applies its power to a[nother] malefic, it will heap evil upon evil; and if a benefic [applies its power] to a benefic, it will add goodness to goodness. If a malefic [applies it power] to a benefic, the thing will change from evil to good; and if a benefic [applies its power] to a malefic, it will change from good to evil.

27) Every planet, whether benefic or malefic, if it is in its domicile or domicile of exaltation,[2] will always indicate good.

28) Any planet that is [positioned] at the beginning of the sign is considered weak until it reaches 5 degrees away from it. Likewise, if the planet is away from a house [before the cusp] by less than 5 degrees, then it is considered [with]in the force of the house, but if more [than that], it is falling [away] from the force of the house.

29) Any planet [positioned] at the beginning of the house up to 15 degrees has great power.

---

[1] That is, the benefic's goodness. [RH]

[2] "Domicile of exaltation" = sign of exaltation. [RH]

30) If a planet is in one of the poles of a malefic[1] and separating [even] by one degree, it will cause fear but will not manifest [the harm] in action.

31) When a planet is in its fall, it indicates worry, distress, and hardship.

32) A retrograde planet indicates antagonism and the destruction of anything that is contemplated.

33) A planet in its first station is like a person who does not know what to do and the result is unfavorable. If it is in its second station, it is like a person hoping for something and not losing one's hope.

34) When the planet is slowing down in its motion, the thing [inquired about] will be delayed whether good or bad, and if Jupiter or Saturn are in the changeable signs, they will expedite it.

35) When a planet is at the end of a sign, it loses its strength and all of its power is in the sign it will enter [next]. If the planet is at the 29th degree of the sign, its influence is still in the sign it is in, because within 3 degrees the planet has influence in the degree it is in, one degree before and one degree after.

36) When a planet is about to join another but before they join the second one enters the following sign, yet the first planet chases it and catches up with it in that place, while no other planet joins with it prior to this, the thing inquired about will be perfected after despair.

37) If a planet aspects the other planet which had moved from its place before the lighter planet caught up with it, it will not harm since the

---

[1] It is not clear whether this means conjunction, square or opposition to the malefic, or whether some text is missing in the sentence. [Additional by RH] The Latin contains nothing to add to this. It is probable that the text is correct as it stands and that it does mean a square or opposition to a malefic.

aspect does not cancel the power of the bodies.[1]

38) If a planet is in a sign of the same [elemental] nature, its strength increases; and if in the opposite nature, it weakens. [An example is] such as Saturn in a cold and dry domicile.

39) A received planet if benefic, then its strength increases, and if malefic, then its evil decreases.

40) When a planet is not in one of its places of rulership and it is in the 6th house or the 12th, it does no good.

41) When a planet is under the light and it is one of the superiors, it has no strength, and so it is for the inferiors. And if they are retrograde, it is the worst.

42) If the significator planet is retrograde and then its motion becomes direct, it indicates some [completion] of the matter. It is the same when the planet is under the light and then comes out from under the light.

43) If a planet is in one of the pits and it is a benefic, its goodness will decrease, and if malefic, its evil will increase.

44) When a benefic planet is in the 8th house, it indicates neither good nor bad, but if it is one of the malefics, it indicates complete harm.

45) When a planet is stationary [about] to retrograde, it indicates no fulfillment of the desired [matter], [and] difficulty and loss. If it is in the direct [motion], it indicates a favorable matter, and its strength [proceeds] straight[forwardly].

46) If the influence of the one-twelfth [position] of the planet [in the sign] is in a good place,[2] it will increase goodness.

---

[1] This aphorism seem like a continuation of the previous one, but it is not entirely clear. [Additional by RH] This relies on the fact a conjunction by body is more powerful than an aspect.

[2] Probably corresponding to a sign that has affinity with the planet, or ruled by it.

47) When a planet is in a fixed sign, it indicates anything that is constant and lasting. If in a changeable [sign], the thing will change, and if in a double-bodied sign, it indicates that some of the matter will last, or it will repeat twice.

48) When the planet receiving the influence is in a bad place, it indicates bad.

49) When the ruler of the rising sign is in its detriment, [the person] does not completely desire the thing inquired about.

50) When the influence [of a planet] is joining [both] benefics and malefics, the nature of the stronger one will manifest.

51) When the ruler of the rising sign confers its power on the ruler of the house [of the matter] inquired about, [the person] will seek the matter with his utmost desire [and will accomplish it], and if the ruler of the matter inquired about confers its power on the ruler of the rising sign, then the matter will be [accomplished] without effort.

52) If there is a planet prohibiting between them, it indicates a person who stands between him and the thing sought after.

53) If the ruler of the rising sign is separating from the ruler of the thing ought after, the person's desire [for it] will go away.

54) If no planet aspects the Moon, this indicates laziness (tardiness).[1]

55) When many planets aspect the Moon, the persons assisting in fulfillment of the matter will be many.

56) When a planet transfers [the influence] from the ruler of the rising sign to the ruler of the thing sought after, the matter will be fulfilled through an intermediary.

57) If the ruler of the rising sign is in a house (sign) of its [own]

---

[1] This is not the same as a void Moon. It simply means that the Moon where it is does not have an aspect. [RH]

triplicity and the [other] rulers of the triplicity aspect it, then his relatives will help him.

58) If a planet is [positioned] in the way we have mentioned as the collection (or transfer) of light, then the matter will be [perfected] after despair.

59) If the significator is in [a position of] conferring of power, it indicates that the matter will be perfected as desired.

60) If the significator is in a position of conferring of rulership, it indicates that the matter will be revealed to another.

61) The testimony of the significator, if [positioned] in conferring of the nature, indicates much joy in the matter.

62) The testimony of the significator, if [positioned] in conferring of two natures, indicates the joy of the person desiring the matter and the person from whom it is requested.

63) The testimony of the significator, if it is [in one of the poles or the adjacent houses],[1] indicates a favorable end for anything that is desired.

64) The testimony of the significator, when in distortion,[2] indicates that [the person] will give up the matter.

65) The testimony of the significator, if in prohibition, indicates that the matter will be destroyed after having been hoped for.

66) The testimony of the significator, if in the returning of the light (to harmful influence), indicates that the inquirer will regret his request.

67) The testimony of the significator, if in cancellation, indicates that there may be circumstances that will destroy the matter.

---

[1] The phrase substitutes the word *yosher* (יושר) and is taken from the corresponding heading in chapter 7.

[2] As in listed in chapter 7, positioned in a cadent house.

68) The testimony of the significator, if in the case [of three planets in one sign],[1] something will happen which will interrupt the inquiry.[2]

69) The testimony of the significator, if in loss, indicates that the inquirer will seek another counsel.

70) The testimony of the significator, if in deprivation of light, indicates a person who destroys the [thing] requested.

71) The testimony of the significator, if in pleasantness, indicates a person who does him a favor.

72) The testimony of the significator, if in recompense, indicates that person will also do favors to others.

73) The testimony of the significators, if in mutual reception, will indicate a thing coming to pass which had not been expected.

74) The testimony of the significator, if in generosity, indicates that the inquirer and the quesited will love each other.

75) The testimony of the significator, if in similitude, indicates every good thing that is possible.

76) The testimony of the significator, if in the evil middle position (besiegement), indicates prison and torture, and if in the good one (between benefics), it indicates maximum benefit.

77) The testimony of the significator, if in rulership, indicates high rank.

78) For the seven planets, when in their places (domiciles), [the rule is] that when a planet is in its domicile, it is like a person in his home.

79) A planet in its exaltation is like a person at his greatest rank.

---

[1] The phrase is taken from the appropriate heading in chapter 7.

[2] It is not clear whether the matter inquired about will be interrupted or the astrological inquiry.

80) A planets in its bound is like a person in his residence.[1]

81) A planet in its triplicity is like a person with his relatives.

82) A planet in its face is like a person with [fine] ornaments and clothing.

83) A planet in its elevation is like a person on his horse.

84) A planet in its similitude is like a person [dealing] in a proper matter.

85) A planet in its dissimilitude is like person [dealing] in an improper matter.

86) A planet in its house of detriment is like a person fighting with himself.

87) A planet in a place where it has no dominion is like a person who is not in his own country.

88) A planet in its house of fall is like a person who has fallen from his great position.

89) A planet under the light of the Sun is like a person in prison.

90) A combust planet is like a dying person.

91) A planet about to turn retrograde is like a frightened person, fearing adversities that are coming to him.

92) A retrograde planet is like a rebellious and defiant person.

93) A planet in its second station is like a person hoping for good [circumstances].

94) A planet slowing down is like a person who is exhausted and has

---

[1] The Latin has 'seat'. [RH]

no strength to walk.

95) A planet swift in its motion is like a young man running.

96) An oriental planet is like person happy in the fulfillment of his desire.

97) An occidental planet is like a lazy man.

98) A planet joined with the Sun is like a person sitting with the king in one chair.

99) An aspecting planet is like a person looking for [a thing] he desires.

100) A separating planet is like a person changing his mind about the matter.

101) A planet in the pole is like person staying in his place.

102) A planet in the adjacent [house] is like person who is hoping.

103) A planet in a falling house is like person going away from his [own] place.

104) Conjoining planets are like two people joining each other.

105) Planets that are aspecting [each other] by a sextile are like two people seeking each other's friendship.

106) When planets aspect each other by a trine, they are like two people of the same nature.

107) When planets are in quartile aspect, they are like two people, each one seeking authority for himself.

108) When planets are in opposition aspect, they are like two people fiercely fighting with each other.

109) A planet in the rising sign is like the newborn that has come out of its mother's womb, or the matter of the moment.

110) A planet in the 2nd house is like a person in the house of his assistants.

111) A planet in the 3rd house is like a person visiting his brothers.

112) A planet in the 4th house is like a person in the house of his forebears or on his land.

113) A planet in the 5th house is like a person in his business and his pleasure.

114) A planet in the 6th house is like a weak man running.

115) A planet in the 7th house is like a person ready for war.

116) A planet in the 8th house is like a person stricken with horror and fear.

117) A planet in the 9th house is like a person exiled from his place or like a person removed from his high position.

118) A planet in the 10th house is like person in his authority, dignity, and profession.

119) A planet in the 11th house is like a person in the house of his friends.

120) A planet in the 12th house is like a person in prison.

Chapter Nine[1]

The ninth chapter [deals] with the lots of the planets and the [lots of the] houses, which are ninety seven.

The Lots of the Planets[2]

The Lot of the Moon. Subtract the position of the Sun in its sign in even[3] degrees from the position of the Moon in her sign, and add the difference to the rising degree. The place where this Lot falls is called the Lot of Good Fortune. You will do so if the nativity is by day. If by night, subtract the position of the Moon from the Sun and add the difference to the rising degree, and there is the Lot of Good Fortune. This is the opinion of the ancients but Ptolemy disagrees with them for he says that we should always subtract the position of the Sun from the Moon whether by day or by night. He is right[4] because as the relation between the rising sign and the Sun so is the relation between the Lot of Good Fortune and the Moon, and therefore it is called the rising sign of the Moon. The Hindu astrologer Masha'allah[5] said in his *Book of Experiments*[6] that the Lot of Mystery[7] (Spirit) by night is stronger than

---

[1] Heading not in original text.

[2] Heading not in the original. [RH]

[3] Ecliptical, not equatorial. See his reasoning in the last paragraph in this chapter. Ibn Ezra repeats this argument in his *Book of Reasons*.

[4] Despite the logic of Ibn Ezra's position, his is a minority opinion. [RH]

[5] Masha'allah was not an Hindu astrologer. In fact he was a Jew as was Ibn Ezra. It makes one wonder how many other ascriptions to Hindus in this text should be questioned. [RH]

[6] I have no idea which of the known works of Masha'allah, if any, this may be. [RH]

[7] The Hebrew word is *ta'aluma* (תעלומה) which means mystery, also implying something hidden or disappearing. This is known now as the Lot of Spirit. [Additional by RH] The Lot of Spirit has a number of other names which may seem strange at first but which serve to elucidate this otherwise unclearly defined point. The central philosophical fact that lies at the basis of this is that many components of the chart can be classified as relating to the material substance out of which something is made, or the formal principle according to which it is made. This has to do with the Aristotelian doctrine of the four causes, material, formal, final, and efficient. Also according to Aristotelian

the Lot of Good Fortune, thus going back to Ptolemy's opinion without noticing it.[1] The Lot of Good Fortune indicates the body, life, wealth, success, good reputation, the beginning of all actions, and whatever is on the mind of the person.

The Lot of the Sun. Subtract the position of the Moon by day from the position of the Sun and add the difference to the rising degree, and so you will find the place of the Lot which is called the Lot of Mystery (Spirit). If the [person is] born at night, then subtract the position of the Sun from the position of the Moon and add the difference to the rising degree and you will find the Lot; and this is the opinion of the ancients. Ptolemy says[2] that it is [calculated in] the same by day and by night, and that is the truth. This Lot indicates the soul, worship of God, and anything that is arcane and exalted.

The Lot of Saturn. Take the distance between the position of Saturn and the degree of the Lot of Good Fortune and add to the rising degree, and that is the Lot if [the person is] born by day; and if by night, it is the reverse. This Lot indicates depth of thought, working the soil (farming), loss, theft, poverty, prison, captivity, and death.

---

doctrine, living things have as their *final* cause (that toward which they grow or evolve) the manifestation of their *formal* cause. Thus the material substance of a living being is intended to grow toward (as its final cause) the most nearly perfect possible manifestation of its form or essence. This linkage between formal and final cause makes it so that anything that has to do with form in astrology is also linked to its future development.

The Lots of Fortune and Spirit are one of these material-formal pairings that one finds in astrology with Fortune pertaining to the material aspect of one's being and Spirit pertaining to the formal-final aspect of one's being. Thus we find that the Lot of Spirit was also called the Lot of Future Things (*Pars Futurorum*). By the same token it may have also been considered to be mysterious precisely because it was linked to an unknowable future, a future in which the perfect (or at least nearly so) form of something came to be manifest as its final cause, hence the name Lot of Mystery.

[1] By night the part of Spirit is Asc + Moon - Sun.

[2] Actually Ptolemy never mentions the Lot of Spirit. The Lot of Fortune is the only one he mentions. This is part of the problem of the Ptolemaic doctrine of the Lot of Fortune. [RH]

The Lot of Jupiter. Take by day the distance between the Lot of Mystery (Spirit) and Jupiter, and add it to the rising [degree], and that is the Lot. By night it is the reverse. This Lot indicates truth, benevolence, wisdom, honor, good reputation, and money.

The Lot of Mars. Take by day the distance between it and the Lot of Good Fortune and add that to the rising degree, and by night the reverse, and that is the Lot. It indicates force, might, anger, swiftness and treachery.

The Lot of Venus. Take by day the distance between the Lot of Good Fortune and the Lot of Mystery (Spirit), and by night the reverse, and add that to the rising degree, and that is the Lot. It indicates love, joy, pleasure, food, drink, lust and intercourse.

The Lot of Mercury. Take by day the distance between the Lot of Mystery (Spirit) and the Lot of Good Fortune, and by night the reverse, and add that to the rising degree. This Lot indicates poverty, enmity and hostility,[1] negotiations, mathematics, and knowledge.

These are the lots of the houses:

The First House has three lots.

One is the *Lot of Life*, and it is computed by day from the distance between Jupiter and Saturn, and by night the reverse, and cast from the rising degree.

The second which is the *Lot of Stature and Beauty* is taken by day from the Lot of Good Fortune to the Lot of Mystery (Spirit), and by night the reverse, and cast from the rising [degree]. This Lot is like that of Venus.

The third which is the *Lot of Intelligence and Speech* is taken by day from Mercury to Mars, and by night it is the reverse, and cast from the rising [degree].

---

[1] However strange it may seem that this Lot is related to Mercury, we have other sources for this such as Al-Biruni, and Bonatti. [RH]

The Second House has three lots.

One is the *Lot of Wealth* taken by day and by night from the lord of the second house to the beginning of the second house calculated for the latitude of the location, and cast from the rising [degree].

The second is the *Lot of Lending*[1] taken by day and by night from Saturn to Mercury, and cast from the rising [degree].

The third is the *Lot of Finding* [things] taken by day from Mercury to Venus and the reverse by night, and cast from the rising [degree].

The Third House has three lots.

One is the *Lot of Brothers* taken by day and by night from Saturn to Jupiter, and cast from the rising [degree].

The second is the *Lot of the Number of Brothers* taken by day and by night from Mercury to Saturn, and cast from the rising [degree].

The third is the *Lot of the Death of Brothers* taken by day from the Sun to the degree of the Midheaven as computed for the latitude of the Earth[2] and the reverse by night, and cast from the rising [degree].

The Fourth House has seven lots.

One is the *Lot of the Father* taken by day from the Sun to Saturn and the reverse by night, and cast from the rising [degree]. If Saturn is under the light of the Sun, then it is taken by day from the Sun to Jupiter and the reverse by night, and cast from the rising [degree].

The second is the *Lot of the Death of the Father* taken by day from Saturn to Jupiter and the reverse by night, and cast from the rising [degree].

---

[1] The Hebrew word *halva'a* (הלואה) can be interpreted either as lending or borrowing.

[2] The Midheaven is not sensitive to latitude, only to longitude.

The third is the *Lot of the Grandfather* taken by day from the ruler of the house of the Sun to Saturn and the reverse by night, and cast from the rising [degree]. If the Sun is in her own domicile or in one of Saturn's domiciles, then it is taken by day from the Sun to Saturn and the reverse by night, and cast from the rising [degree]; and [in this case] you need not be concerned if it (Saturn) is under the light of the Sun.

The fourth is the *Lot of Lineage* taken by day from Saturn to Mars and the reverse by night, and cast from the rising [degree].

The fifth is the *Lot of Property* (land) taken by day and by night from Saturn to the Moon, and cast from the rising [degree].

The sixth is the *Lot of Cultivation of the Soil* taken by day and by night from Venus to Saturn and cast from the rising [degree].

The seventh is the *Lot of the End*[1] taken by day and by night from Saturn to the ruler of the conjunction of the lights (New Moon) if the native was born in the first half of the [lunar] month, or to the ruler of the opposition (Full Moon) if born in the second half, and cast from the rising [degree].

The Fifth House has five lots.

One is the *Lot of the Child*[2] taken by day from Jupiter to Saturn and the reverse by night, and cast from the rising [degree].

The second is the *Lot for the Time the Child is Born*, whether male or female, taken by day and by night from Mars to Saturn, and cast from the rising [degree].

The third is the *Lot of Male Children* taken by day and by night from the Moon to Jupiter, and cast from the rising [degree].

---

[1] 'End of the matter' or 'end of life'.

[2] The word is *haben* (הבן) which means 'son', but the differentiation between male and female in the next lots suggests that this one is for children in general. [Additional by RH] Other sources confirm this statement.

The fourth is the *Lot of Daughters* taken by and by night from the Moon to Venus, and cast from the rising [degree].

The fifth is the *Lot of the Question Whether Male or Female* taken by day from the ruler of the house of the Sun to the Moon, and cast from the rising [degree].[1]

The Sixth House has three lots.

One is the *Lot of Diseases and Handicap* taken by day from Saturn to Mars and the reverse by night, and cast from the rising [degree].

The second is the *Lot of Slaves* taken by day and by night from Mercury to the Moon, and cast from the rising [degree].

The third is the *Lot of Prison and Captivity* taken by day and by night from the ruler of the house of the Sun to the Sun, and by night from the ruler of the house of the Moon to the Moon, and cast from the Ascendant.

The Seventh House has thirteen lots.

One is the *Lot of Marriage*[2] *for [both] Men and Women* taken by day and by night from Venus to the setting degree, and cast from the Ascendant.

The second is the *Lot of Marriage in the Nativity of Men* according to Enoch taken by day and by night from Saturn to Venus, and cast from the Ascendant, and according to Valens taken by day and by night from the Sun to Venus, and cast from the Ascendant.

The third is the *Lot of the Time of Marriage* taken by day and by night

---

[1] Neither the Hebrew nor the Latin tell us what to do in a night birth. Abu Ma'shar in the *Abbreviation,* chapter 6 [27] tells us to reverse the calculation at night. [RH]

[2] The Hebrew word here is *be'ilut* (בעילות) which also means 'intercourse'. The word *ba'al* (בעל) means 'husband', 'owner', and is also the root of the verb for a man having intercourse with a woman, usually in a marriage.

from the Sun to the Moon, and cast from the Ascendant.

The fourth is the *Lot of Adultery in Marriage* taken by day and by night from the Sun to the Moon, and cast from the degree of Venus, and there is the Lot.[1]

The fifth is the *Lot of Modesty* taken by day and by night from the Moon to Venus, and cast from the Ascendant.

The [sixth is], the *Lot of Prostitution of Women* taken by day and by night from Venus to Saturn, and cast from the Ascendant.

The seventh is the *Lot of Male Adultery* taken by day and by night from the Sun to Venus, and cast from the Ascendant.

The eighth is the *Lot of Female Adultery* taken by day and by night from the Moon to Mars, and cast from the Ascendant.

The ninth is the *Lot of Prostitution of Males* taken by day and by night from the Sun to Venus, and cast from the Ascendant.

The tenth is the *[Lot of] Intercourse* taken by day and by night from the Moon to Mars, and cast from the Ascendant.

The eleventh is the *Lot of [Sexual] Desire* taken by day and by night from the Moon to Mars, and cast from the Ascendant.

The twelfth is the *Lot of Quarrels* taken by day and by night from Mars to Jupiter, and cast from the Ascendant.

The thirteenth is the *Lot of Bridegrooms* taken by day and by night from Saturn to Venus, and cast from the Ascendant.

The Eighth House has five lots.

One is the *Lot of Death* taken by day and by night from the degree of the Moon to the beginning of the eighth house calculated according to

---

[1] An instance of a lot that is not projected from the Ascendant. [RH]

the latitude of the place, and cast from the place of Saturn, and there is the Lot.

The second is the *Lot of the Killing Planet* taken by day from the ruler of the rising degree to the Moon and the reverse by night, and cast from the Ascendant.

The third is the *Lot of the Year of Danger* taken by day and by night from Saturn to the lord of the house [where] the conjunction of the luminaries [occurred before birth] if the nativity was in the first half of the [lunar] month, or to the lord of the house of the opposition (Full Moon) if the nativity was in the latter half of the [lunar] month, and cast from the Ascendant.

The fourth is the *Lot of the Place of Sickness* taken by day from Saturn to Mars and the reverse by night, and cast from the place of Mercury, and there is the Lot.

The fifth is the *Lot of Distress* taken by day from Saturn to Mercury and the reverse by night, and cast from the Ascendant.

The Ninth House has seven lots.

One is the *Lot of Travel* taken by day and by night from the ruler of the ninth house to the beginning of the ninth house calculated according to the latitude of the place, and cast from the Ascendant.

The second is the *Lot of Travel by Water* taken by day from Saturn to 15 degrees of Cancer and the reverse by night, and cast from the Ascendant. If Saturn happens to be at that degree, the Lot will be in the Ascendant.[1]

The third is the *Lot of Humbleness* taken by day from the Moon to Mercury and the reverse by night, and cast from the Ascendant.

The fourth is the *Lot of Wisdom* taken by day from Saturn to Jupiter and the reverse by night, and cast from the place of Mercury, and there

[1] Because then the distance from Saturn to 15° Cancer would be zero.

is the Lot.

The fifth is the *Lot of Knowledge* taken by day from Saturn to Jupiter and the reverse by night, and cast from the Ascendant.

The sixth is the *Lot of Prophecies*[1] taken by day from the Sun to Jupiter and the reverse by night, and cast from the Ascendant.

The seventh is the *Lot of Whether the Matter is True or False* taken by day and by night from Mercury to the Moon, and cast from the Ascendant.

The Tenth House has eleven lots.

One is the *Lot of Kingship* taken by day from Mars to the Moon and the reverse by night, and cast from the Ascendant.

The second is the *Lot of Triumph* taken by day from the Sun to Saturn and the reverse by night, and cast from the Ascendant. If Saturn happens to be under the light of the Sun, then it is taken by day from the Sun to Jupiter and the reverse by night, and cast from the Ascendant.

The third is the *Lot of Counsel*[2] taken by day from Mercury to Mars and the reverse by night, and cast from the Ascendant.

The fourth is the *Lot of Philanthropy* taken by day and by night from Mercury to the Sun, and cast from the Ascendant.

The fifth is the *Lot of Sudden Taking Over of Power* taken by day from Saturn to the Lot of Good Fortune and the reverse by night, and cast from the Ascendant.

---

[1] The word *hagadot* (הגדות) also means 'telling stories'.

[2] The word is *etza* (עצה) which means counsel or advice. It may also mean 'intending or planning to do something'. The latter may be the more correct translation in the context of the 10th house and also because Mercury provides the intelligence while Mars provides the action.

The sixth is the *Lot of Activity (or Profession)* taken by day and by night from Saturn to the Moon, and cast from the Ascendant.

The seventh is the *Lot of Work Done with the Hands* taken by day from Mercury to Venus and the reverse by night, and cast from the Ascendant.

The eighth is the *Lot of the Work (or Action) which Must Be Done* taken by day from the Sun to Jupiter and the reverse by night, and cast from the Ascendant.

The ninth is the *Lot of Commerce (or Business)* taken by day from the Lot of Mystery (Spirit) to the Lot of Good Fortune and the reverse by night, and cast from the Ascendant.

The tenth is the *Lot of Greatness* taken by day from the Sun to the degree of its exaltation (19 Aries) and by night from the Moon to the degree of its exaltation (3 Taurus). If the Sun by day is in the degree of its exaltation or the Moon by night is in the degree of its exaltation, then the Lot will be in the Ascendant.

The eleventh is the *Lot of the Mother* taken by day from Venus to the Moon and the reverse by night, and cast from the Ascendant.

The Eleventh House has ten lots.

One is the *Lot of the Friend* taken by day from the Lot of Good Fortune to the Lot of Mystery (Spirit) and the reverse by night, and cast from the Ascendant.

The second is the *Lot of Being Known Among People* taken by day from the Lot of Good Fortune to the Sun and the reverse by night, and cast from the Ascendant.

The third is the *Lot of Prosperity* taken by day from the Lot of Good Fortune to Jupiter and the reverse by night, and cast from the Ascendant.

The fourth is the *Lot of Hope* taken by day from Saturn to Venus and the reverse by night, and cast from the Ascendant.

The fifth is the *Lot of Abundance at Home* taken by day and by night from the Moon to Mercury, and cast from the Ascendant.

The sixth is the *Lot of Freedom of Soul* taken by day from Mercury to Jupiter and the reverse by night, and cast from the Ascendant.

The seventh is the *Lot of Praise* taken by day from Jupiter to Venus and the reverse by night, and cast from the Ascendant.

The eighth is the *Lot of Desire* taken by day from the Lot of Good Fortune to the Lot of Mystery (Spirit) and the reverse by night, and cast from the Ascendant.

The ninth is the *Lot of Change* taken by day and by night from the Lot of Mystery (Spirit) to Mercury, and cast from the Ascendant.

The tenth is the *Lot of Friends* taken by day and by night from Mercury to the Moon, and cast from the Ascendant.

The Twelfth House has two lots.

One is the *Lot of Enemies* taken by day and by night from the lord of the 12th house to the beginning of the twelfth house as calculated for the latitude of the place, and cast from the Ascendant.

The Second Lot According to Enoch is taken by day from Saturn to Mars and the reverse by night, and cast from the Ascendant.

The total of the lots of the houses is 71.

The Lots That Are by Themselves are nine.

One is the *Lot of the Number of Years of Life*. Look to see if the nativity was in the first half of the [lunar] month then take the distance from the degree of the conjunction of the lights (new Moon) to the degree of the Moon at the time of the nativity and cast it from the Ascendant. If the nativity was in the latter half of the month, then take the distance from the degree of the opposition of the lights (Full Moon) to the degree of the Moon and cast from the Ascendant.

The second is the *Lot of Defect in the Body* taken by day from the Lot of Good Fortune to Mars and the reverse by night, and cast from the Ascendant.

The third is the *Lot of Delay* taken by day and by night from the degree of Mars to the beginning of the 3rd house as calculated for the latitude of the place, and cast from the Ascendant. Enoch said that it is most properly taken from the Lot of Love to Mercury and cast from the place of Mercury, and there is the Lot.

The fourth is the *Lot of Deceit* taken by day from Mercury to the Lot of Mystery (Spirit) and the reverse by night, and cast from the Ascendant.

The fifth is the *Lot of the Place Sought* taken by day and by night from Saturn to Mars, and cast from the place of Mercury.

The sixth is the *Lot of Following* taken [by day] from Mars to the Sun and the reverse by night, and cast from the Ascendant.

The seventh is the *Lot of Truth* taken by day from Mercury to Mars and the reverse by night, and cast from the Ascendant.

The eighth is the *Lot of Might* taken by day from Saturn to the Moon and the reverse by night, and cast from the Ascendant.

The ninth is the *Lot of Killing* taken by day from the ruler of the Ascendant to the Moon and the reverse by night, and cast from the Ascendant.

These are the lots that are necessary at the time (return) of the Year of the Great Conjunction[1] in order to know the affairs of kings and how long they will reign.

Of these is the lot called the *Lot of Kingdom* necessary for the years of the world (history) taken by day from Mars to the Moon, and cast from the Ascendant in the year of the Great Conjunction, which indicates

[1] Jupiter and Saturn.

displacement of the nation.[1] Some extract this Lot in another way; [it is] taken from the Ascendant [[in the annual revolution of the year of the [Great] Conjunction to the degree of the conjunction, and cast from the Ascendant]] in the annual revolution [of that year].

Some extract it in a different way; [it is] taken from [the degree of the] Midheaven vis-a-vis the place of the Sun to the Midheaven in the annual revolution, and cast from the place of Jupiter.

The *Lot of the Number of Days [of Reign] of the Kings* is taken in the year of the king's ascent to the throne from the Sun[2] to 15 degrees of Leo and cast from the place of the Moon. It is also taken from the place of the Moon to 15 degrees of Cancer and cast from the place of the Sun.

Know that these two lots always fall in houses (domiciles) of the same planet,[3] and if the lot falls in the house of one luminary, the other lot falls in the house of the other luminary.

Another *Lot in the Year of the King's Ascendancy* is taken by day from Jupiter to Saturn and the reverse by night, and cast from the Ascendant in the annual revolution. If Jupiter happens to be in a double-bodied

---

[1] This lot and the next one are both based on the Aries ingress chart. [RH]

[2] It is not clear whether he means the Sun at the Vernal Equinox, which is always 0 degrees Aries, or the Sun in chart of the ascent to the throne.

[3] Examples: Formulas First Lot = ☽+15°♌-☉
Second Lot = ☉+15°♋-☽

Assume ☉15°♉, ☽15°♊.
First lot: 15°♌-☉=90°; ☽+90°=**15°♍** = Lot
Second lot: 15°♋-☽=30°;☉+30°=**15°♊** = Lot
Both lots are ruled by Mercury.

Assume ☉15°♓,☉15°♑.
First lot: 15°♌-☉=150°; ☽+150°=**15°♊** = Lot
Second lot: 15°♋-☽=180°;☉+180°=**15°♍** = Lot
Both lots are ruled by Mercury.

Assume ☉15°♉,☽15°♐.
First lot: 15°♌-☉= 90°; ☽+ 90°=**15°♓** = Lot
Second lot: 15°♋-☽=210°;☉+210°=**15°♐** = Lot
Both lots are ruled by Jupiter.

sign, and the revolution is diurnal, and Jupiter is in a cadent house, then it is taken from Saturn to Jupiter with an addition of 30 degrees, and cast from the Ascendant. If Saturn happens to be opposite Jupiter and they are both in cadent houses, then half of the distance between them is taken and cast from the Ascendant. If Jupiter is in its house of dignity and the revolution is nocturnal, then it is taken from him to Saturn and cast from the Ascendant.

The Great Lots are two. One is when you look in the year of the king's ascendancy [to see] where the place of the conjunction of the triplicity[1] which took place prior to the king's ascendancy has reached by giving every 30 degrees one year and every 2½ degrees one month. When you know the degree arrived at by the calculation, from there you will determine the first lot. [Now you need to] know the revolution in the year of the king's ascendancy, and look to Saturn or Jupiter, whichever is oriental of the Sun, and take the distance between it at the beginning of the annual revolution, and the degree I have mentioned, as from there you will determine the Lot, and cast that difference from the Ascendant at the beginning of the annual revolution, and there is the Lot.

The second lot is when you look to Jupiter or Saturn, whichever is occidental of the Sun at the moment of the annual revolution, and take the distance between it and the degree I have mentioned from which you determined the first lot, and project the difference from the Ascendant of the annual revolution.

The Lot of Rain. Take by day the distance between the degree of the conjunction of the luminaries (New Moon) to the degree of the Moon, and by night from the Moon to the degree of that conjunction and cast from the rising sign in the morning or in the evening,[2] and the place where the calculation is complete there is the Lot. Now, if any of the stars (planets) is with this Lot in one of the angles, know that when the

---

[1] That is, the first conjunction *within* the triplicity.

[2] The precise meaning of the phrase "in the morning or in the evening" is not clear. The Latin says exactly the same thing. We know that the formula for the Lot changes from day to night. One possible interpretation is that this Lot should be cast only at sunrise or sunset but that is about the only sense that can be made of it. Unfortunately we have no other reference to such a lot with a formula anything like this one. [RH]

Moon reaches the place of that star, that is when it will rain. This will be [true] in every country that is ruled by the sign where that Moon is.

Enoch said that it is taken every day from the place of the Sun to the place of Saturn and cast, in the morning, from the place of the Moon, and that is the Lot, and that has been tried. If the Lot falls in one of Saturn's houses, there will be cold, and if in one of Jupiter's houses, there will be winds, and if in one of Mars' houses, there will be heat, and if in one of Venus's houses, there will be rain or fog, and if in one of Mercury's houses, there will be strong winds, and if in the Moon's house, there will be cloud or rain, and if in the Sun's house, the air will be pure.

The Lot of Whether a Matter Will Come To Pass. Take the distance between the Sun and the ruler of the hour, and multiply the result by the whole crooked hours[1] that passed of the day or the night, and save the result. Then take the ascension of the sign where the ruler of the hour is for the latitude of the location, and multiply by 70, and divide the result by 12. Add the result to the saved [number], and cast the total from the place of the ruler of the hour, and there is the Lot.

These are the **Lots, in Every Revolution, to Know All the Things That Will Become More Expensive or Cheap**, and this is what you do.

Look to the place of the lot to see which is its ruler, and whether it is retrograde, or combust, or in a house falling from the angles (cadent) according to the Ascendant in the annual revolution, then whatever that lot signifies will be cheap. If the ruler of the house of the lot is in its power or in the Midheaven, then it will become expensive. When the ruler of the house of the lot reaches its house of dignity, then it will increase in price, and if [it reaches] its house of debility, it is the reverse. If one of the benefic planets aspects the lot, the commodity that the lot indicates will increase, and if the malefics aspect it, then it will decrease and be destroyed.

The *Lot of Water* is taken from The Moon to Venus.
The *Lot of Wheat* is taken from the Sun to Mars.
The *Lot of Barley* is from the Moon to Jupiter.

---

[1] Uneven, temporal hours.

The *Lot of Peas* is from Venus to the Sun.
The *Lot of Lentils* is from Mars to Saturn.
The *Lot of Beans* is from Saturn to Mars.
The *Lot of Butter* is from the Sun to Venus.
The *Lot of Honey* is from the Moon to the Sun.
The *Lot of Rice* is from Jupiter to Saturn.
The *Lot of Olives* is from Mercury to the Moon.
The *Lot of Grapes* is from Saturn to Venus.
The *Lot of Cotton* is from Mercury to Venus.
The *Lot of Watermelons* is from Mercury to Saturn.
The *Lot of Sour Foods* is from Saturn to Mars.
The *Lot of Sweet Foods* is from the Sun to Venus.
The *Lot of Spicy Foods* is from Mars to Saturn.
The *Lot of the Bitter Foods* is from Mercury to Saturn.
The *Lot of Medicinal Herbs that are Sour and Laxative* is from Saturn to [[Jupiter]].
The *Lot of Medicinal Herbs that are Salty* is from Mars to the Moon.
The *Lot of Deadly Poison* is from the North Node to Saturn.

All these lots are cast from the Ascendant.

These lots that we have mentioned, were extracted by the ancients for two reasons. One is because the planets' co-mixture with each other by conjunction or aspect is in many ways, and from their co-mixture comes forth good and bad in all locations and at every moment. [Therefore,] the more manifest power is of the lot that is taken from two planets that signify the same thing, such as the Sun and Saturn, both of which signify the fathers. Therefore, we need to know the distance between them at any moment in order to investigate the matter of fathers. The second reason is that every thing that is indicated by the stars needs two or three witnesses. Now, the testimony can be in doubt as the planet may be one of the nocturnal planets and the other is of the diurnal planets, then the one will be stronger than the other, or, when one indicates the beginning of the matter and the other indicates its end. Therefore we need to extract the lots.

The lot is [based] on three components. Two of them are fixed and the third is the variable. One fixed [component] is that from which the lot is taken, the second is the one to which it is taken, and the third is the Ascendant and that is the variable one since it changes every moment. We must cast these lots from the Ascendant since it indicates the beginning of actions. They are also cast from the beginning of the house that signifies the matter.

These lots are extracted in even degrees,[1] because the planets move in the wheel of the zodiac, and when it is said that a planet is in a certain sign and a certain degree [[and the Ascendant]] is in a certain sign, they are [stated] in even degrees.[2] Therefore, the lots were extracted in [[even]] degrees. Only the degrees of the [signs'] ascension [are uneven] as they are the degrees of the superior wheel which is above the wheel of the zodiac, for it turns the wheel of the zodiac on its two axes. Therefore all aspects are considered in even degrees in the wheel of the zodiac as was tried by the ancients in nativities, and inquiries, and elections; and when the Moon applies her influence to one of the planets in even degrees through the power of the aspects, they saw that their judgments were always true by the quartile, the trine, and the sextile aspect. We have tried this in the same way and succeeded.

---

[1] As opposed to uneven degrees if we measure by oblique ascension.

[2] Some more modern astrologers, such as James Wilson have insisted that the Lot of Fortune should be erected in oblique ascension. Obviously Ibn Ezra disagrees. [RH]

Chapter Ten[1]

The tenth chapter [deals with] aspects and directions.

Know that [for] the aspects the directions to them are also in two ways. One way is by the aspect according to the computation of the ascension of the signs in each and every country. The second way is by the number of degrees in the signs, which are the same in each and every country.

If the planet is in the Ascendant, then you should direct it according to the ascension of the signs for that place. If it is at the beginning of the 7th house, you should direct it by [using] its exchanged degree, which is its opposite degree.[2] If it is at the beginning of the line of the Midheaven or the beginning of the 4th house, then you should direct it by the ascension of the signs in the right [ascension] wheel, and that is the opinion of Enoch.[3]

The second way according to all astrologers is that you should direct every planet or lot which you need to direct in whatever place it is in the wheel in even degrees in the wheel of the zodiac, and not in the right [ascension] wheel. This has been tried endless times and was found true. Thus, the opposition aspect will always be 180 degrees, the trine aspect 120 degrees, the quartile aspect 90 degrees, and the sextile aspect 60 degrees, and they are all even degrees. This is true if the planet has no distance from the ecliptic, but if it has latitude, it will be somewhat detracted. That is what Albatani[4] said in his book, and all the ancients [and those] who came after them.

The reason for these directions is that when you direct a planet or a degree to the body of a planet, or to the aspect of its light, [you will be able] to know how many years there are between them. By directions there shall be known all the good and the bad that shall befall kings, and the conquest of the kingdom of one nation by another, and changes that occur in the world in general, and in particular from evil to good and from good to evil.

The directions are [figured] in five ways. [One,] in order to know

---

[1] Heading not in original text. [RH]

[2] This is equivalent to using the oblique *descension* of the degree itself. [RH]

[3] And of Ptolemy.

[4] Known in Latin as Albategnius. [RH]

the general matters of the world, such as the deluge and the drought that destroy the world, or the affairs of countries such as wars and new laws. [For this purpose] the world is under the influence of one sign for 1000 years, and the part for one year in this first direction is 1 [minute] and 48 seconds.[1]

The second [type of] direction is under the direction of the thousands, and it indicates all that happens to every people and every region. [In this,] the world is under the influence of one sign for 100 years, and the part for one year in this direction is 18 minutes.

The third way is under the direction of the hundreds, and it indicates all that occurs in every country and every family. The world is under the influence of one sign for 10 years [in this direction], and so is every person under the influence of one sign for 10 years until 120. The part for one year in this direction is 3 degrees.

The fourth way is the direction called Al-fardar, directed as follows in the affairs of the world. It is begun from any planet whose ancient[2] house of dignity is opposite of the sign of Aries. For those born by day it is begun from the Sun and by night from the Moon. The total figure for the time of the Al-fardar is 75.[3]

The fifth way is the direction of the single [years] in worldly affairs, to know what will happen every year. In the nativity too, every year is under the influence of one sign, and it is called the sign that completes a revolution. This direction is completed in 12 years, so the part for one year is 30 degrees, which are the degrees of one sign. This is [also] how you should direct the great conjunction when Saturn and

---

[1] This system is known in the work of Abu Ma'shar as the *Intiha'at,* singular *Intiha.* The mighty *Intiha* is the one described in the preceding paragraph. The next paragraphs describe the big, and middle *Intiha'at* respectively. The small *Intiha* is described a few paragraphs down along with what is clearly a profection of one sign per year. Could this system be in part a logical extension upwards of the profections? See David Pingree, *The Thousands of Abu Ma'shar*, London: Warburg Institute, 1968, p. 60.

[2] The word here is *kadmon* (קדמון) which usually means 'ancient'. In this context it might also mean preceding. The root of the word (קדם) is also used for 'east' (as east is before west), and also in words that denote 'front' or 'ahead'.

[3] This system is a bit different from the firdar reported by Al-Biruni and other medieval authorities in that the planets do not follow the Chaldean order. [RH]

Jupiter move from one triplicity to another until they return to their first place, which takes place after 960 years. The part for one year [in this case] is 22 minutes and 16 seconds and 5 thirds.[1] You should also direct the middle conjunction, which is the change of the above from one triplicity to another, and that takes place in 240 years. The part for one year is 1 degree and 29 minutes and 4 seconds. You should also direct the minor conjunction when the [above] mentioned change from one sign to another within the signs of the [same] triplicity which takes place every 20 years approximately. The part for one year is 28 degrees approximately.

There is one more direction in the nativity of a person and in the annual revolution for the world, and that is directing from the Ascendant to the body of a planet or to its aspect at a certain sign, or to a certain degree, for each year one degree. The direction of the lots is in the opposite [sequence] of the signs, as Ptolemy mentioned in the *Book of Fruit.*[2]

Here end the ten chapters. Praise to the Creator of all created who has given the author strength and he completed [them] in the month of Tamuz the year 4908.

Final, Final, Praise to God Eternal.

---

[1] A division below seconds. The idea is equating the period of 960 years with one revolution of the signs, thus 360/960 yields 0°22' 30" per one year.

[2] This is the actual traditional name of the *Centiloquy* attributed (incorrectly) to Ptolemy. However, I can find no reference to the direction of lots in the *Centiloquy*.

www.ingramcontent.com/pod-product-compliance
Lightning Source LLC
LaVergne TN
LVHW090950080826
845145LV00003B/955